IntechOpen

Supporting open minds since 2005

Remote Sensing
http://dx.doi.org/10.5772/intechopen.87829
Edited by Andrew Hammond and Patrick Keleher

Contributors
Maria Lazaridou, Ioannis Kotaridis, Dibyendu Dutta, Akanksha Balha, Prabir Kumar Das, Pragyan Jain, Libeesh Lukose, Wasim Akram, Eduardo Landulfo, Alexandre Cacheffo, Pablo Ristori, Juan-Carlos Antuña-Marrero, Paulo Batista, Alexandre Pimenta, Maria Paulete Pereira Martins, Vania Andrioli, Chi Wang, Jiyao Xu, Boris Barja González, Diego Alves Gouveia, Eduardo Quel, Felix Zamorano, Gregori Arruda Moreira, Henrique Barbosa, Jacobo Salvador, Lidia Otero, Marcos Andrade, Ricardo Forno, John Reina, Estivén Sanchez, Alexandre Yoshida, Fabio Lopes, Facundo Casasola, Clodomira Pereyra, Maria Prietro, Silvina Brusca, Cristina Sousa, Daniel Santos, Ediclê Duarte, Elena Montilla-Rosero, Judith Hoelzemann, Marcos Araujo, Jonatan Da Silva, Juan Pallotta, Eliam Wolfram, Facundo Orte, Renata Sammara, Antonio Arleques, Sabelo Nick Dlamini, Ehsan Atazadeh, Benjamin Arias-Perez, Antonio Arleques Gomes, Jonatan João da Silva, Facundo Ismael Casasola, Jacobo Omar Salvador, Lidia Ana Otero, Juan Carlos Antuña-Marrero, Anderson Guimarães Guedes, Daniel Camilo Fortunato dos Santos Oliveira, Marcos Paulo Araújo da Silva, Mostafa Mahdavifard

First published in London, United Kingdom, 2021 by IntechOpen
IntechOpen is the global imprint of INTECHOPEN LIMITED, registered in England and Wales, registration number: 11086078, 5 Princes Gate Court, London, SW7 2QJ, United Kingdom
Printed in Croatia

British Library Cataloguing-in-Publication Data
A catalogue record for this book is available from the British Library

Additional hard and PDF copies can be obtained from orders@intechopen.com

Remote Sensing
Edited by Andrew Hammond and Patrick Keleher
p. cm.
Print ISBN 978-1-83880-978-2
Online ISBN 978-1-83880-979-9
eBook (PDF) ISBN 978-1-83880-980-5

Meet the editors

Dr. Hammond is a Senior Lecturer in Geoscience within the School of Engineering & Technology at Central Queensland University's Rockhampton Campus, Central Queensland, Australia. He holds undergraduate degrees in Geology and Geography from the University of Tasmania and postgraduate degrees, Master of Applied Science in Pedology from the University of Canterbury, and Ph.D. in Earth Science from Massey University, both in New Zealand. His research interests are basin tectonics and stratigraphy, sedimentology, soil/regolith geology, hydrogeology, geohazards, mining education, and the use and interpretation of geospatial imagery for environmental assessment and land resource management. He has over 30 years' geoscience consultancy, research, and teaching experience having worked in a range of geological landscapes within Australia and New Zealand. He held senior research scientist positions in State (Tasmania) and Australian Government agencies and as a researcher in Queensland university research centers.

Dr. Patrick Keleher has qualifications in physics, mathematics, education, robotics and management psychology. His career has focused on developing and delivering innovative curriculum in the areas of physics, mathematics, maintenance engineering, chemistry, Futures Studies, control systems and environmental science. Dr. Keleher has conducted research, including the supervision of undergraduate and postgraduate candidates, into the areas of maintenance engineering, environmental monitoring, management practices and work-integrated learning. He has served as a visiting professor and/or a keynote speaker at the Open University, Milton Keynes, United Kingdom, the Duale Hochschule Baden-Württemberg Co-operative University (DHBW), Ravensburg, Germany and at Beijing University of Technology, Beijing, China. Delivering workshops and keynotes on such diverse topics as Work Integrated Learning, Futures Studies, Corporate Governance, Intercultural Management and Engineering Professional Standards. Dr. Keleher has been the recipient of a CQUniversity Vice Chancellor's Awards for Outstanding Contributions to Learning and Teaching for the design and delivery of two undergraduate first-year physics units. He has been acknowledged as a distinguished scholar, as a Fellow of the Practice-Based Professional Learning Centre for Excellence in Teaching and Learning (CETL), at The Open University, Milton Keynes, United Kingdom.

Contents

Preface

This Edited Volume is a collection of reviewed and relevant research chapters, concerning the developments within the Remote Sensing. The book includes scholarly contributions by various authors and edited by a group of experts pertinent to this subject field. Each contribution comes as a separate chapter complete in itself but directly related to the book's topics and objectives.

Chapters included in this book are:

"Lidar Observations in South America. Part I - Mesosphere and Stratosphere"

"Lidar Observations in South America. Part II - Troposphere"

"Application of Remote Sensing in Natural Sciences"

"Assessment of Ecological Disturbance Caused by Flood and Fire in Assam Forests, India, Using MODIS Time Series Data of 2001-2011"

"Delineation of Open-Pit Mining Boundaries on Multispectral Imagery"

"Stereoscopic Precision of the Large Format Digital Cameras"

"Remote Sensing Applications in Disease Mapping"

The target audience comprises scholars and specialists in the field.

IntechOpen

Chapter 1

Lidar Observations in South America. Part I - Mesosphere and Stratosphere

Eduardo Landulfo, Alexandre Cacheffo, Alexandre Calzavara Yoshida, Antonio Arleques Gomes, Fábio Juliano da Silva Lopes, Gregori de Arruda Moreira, Jonatan João da Silva, Vania Andrioli, Alexandre Pimenta, Chi Wang, Jiyao Xu, Maria Paulete Pereira Martins, Paulo Batista, Henrique de Melo Jorge Barbosa, Diego Alves Gouveia, Boris Barja González, Felix Zamorano, Eduardo Quel, Clodomyra Pereira, Elian Wolfram, Facundo Ismael Casasola, Facundo Orte, Jacobo Omar Salvador, Juan Vicente Pallotta, Lidia Ana Otero, Maria Prieto, Pablo Roberto Ristori, Silvina Brusca, John Henry Reina Estupiñan, Estiven Sanchez Barrera, Juan Carlos Antuña-Marrero, Ricardo Forno, Marcos Andrade, Judith Johanna Hoelzemann, Anderson Guimarães Guedes, Cristina Tobler Sousa, Daniel Camilo Fortunato dos Santos Oliveira, Ediclê de Souza Fernandes Duarte, Marcos Paulo Araújo da Silva and Renata Sammara da Silva Santos

Abstract

South America covers a large area of the globe and plays a fundamental function in its climate change, geographical features, and natural resources. However, it still is a developing area, and natural resource management and energy production are far from a sustainable framework, impacting the air quality of the area and needs much improvement in monitoring. There are significant activities regarding laser remote sensing of the atmosphere at different levels for different purposes. Among these activities, we can mention the mesospheric probing of sodium measurements and stratospheric monitoring of ozone, and the study of wind and gravity waves. Some of these activities are long-lasting and count on the support from the Latin American Lidar Network (LALINET). We intend to pinpoint the most significant scientific achievements and show the potential of carrying out remote sensing activities in the continent and show its correlations with other earth science connections and synergies. In Part I of this chapter, we will present an overview and significant results of lidar observations in the mesosphere and stratosphere. Part II will be dedicated to tropospheric observations.

Keywords: lidar, LALINET, aerosols, atmospheric sciences, remote sensing, air quality, environment

1. Introduction

Currently, the world's leading authority on global warming issues is the Intergovernmental Panel on Climate Change (IPCC). The IPCC is a scientific-political organization, created in 1988 by the United Nations (UN), and received the Nobel Peace Prize in 2007 [1, 2]. Since its foundation, the IPCC has issued five reports (Assessment Reports), the first in 1990, the next ones in 1995, 2001, 2007, and 2014. The next report of IPCC is expected for the year 2022. The IPCC reports have reinforced, with growing evidence, that human influence on Earth's climate is incontestable and that the terrestrial climate system's warming is evident [2].

Aerosols, in particular, can alter the most diverse atmospheric processes, significantly affecting weather and climate. For example, they can absorb or scatter specific solar radiation wavelengths and radiation reflected by the Earth's surface [3]. They can also modify the albedo (ability to reflect solar radiation on a given surface) and the lifetime of clouds [4]. A decrease in the albedo of clouds, for example, can lead to less reflection of radiation from the Sun, contributing to possible global warming effects. In this context, it is expected that the aerosol climatological behavior in the Earth's atmosphere and its influence on climate change processes are of paramount importance.

The World Meteorological Organization (WMO) has encouraged the creation and expansion of networks aimed at atmospheric observations, and ground-based lidar networks have acquired great importance, both for atmospheric monitoring and research. Thus, regional lidar networks' development to research the most diverse atmospheric configurations is strategic. The main fields where ground-based lidar measurements can be applied include [5, 6] atmospheric aerosol optical properties, urban aerosols and pollution, dust and biomass burning transportation, and cloud impacts on climate, planetary boundary layer dynamics, and processes of satellite data validation.

In terms of atmospheric structure, ground base lidars cover from the mesosphere down to the troposphere, through the stratosphere, and inspect each atmospheric layer in question. Under this perspective, laser radars' operation began in the early '70s by observing stratospheric aerosols in Brazil and continued with sodium atoms (Na) concentration in the mesosphere. The stratospheric aerosols and ozone studies followed some years later in Argentina [7] and the late '80s in Cuba. By the late '90s and early 2000, the introduction of the lidar for tropospheric studies began.

We intend to summarize the most significant scientific achievements and developments related to ground-based Lidar remote sensing in South America in the next sections. LALINET's most recent efforts in establishing standard protocols of system configurations, quality assurance, measurements, and data processing also will be approached [7–11]. The chapter organization should first follow the studies performed in the mesosphere, followed by the work devoted to the stratosphere, and then we should show the studies related to the troposphere. These sections will be distributed over many specific studies regarding the scientific drives and methodologies employed.

2. Lidar remote sensing in Latin America: LALINET

The South American continent, encompassing 42% of the Americas, is a region that shelters the most remarkable ecosystems. Among these, we can cite the Amazon Rainforest, which is the largest tropical forest in the world, the Pantanal (or Chaco), one of the UNESCO World Heritage Sites [12], and the Andes, the most extensive mountain chain in the world, and which hold a plethora of active and inactive volcanoes, extending from Venezuela to Patagonia, crossing all the continent from north to south. Patagonia, the continent's southern region, presents many plants and wildlife, mostly endemic. It also houses another UNESCO World Heritage Site: The National Park Los Glaciares, in Santa Cruz, Argentina [12].

Figure 1.
Schematic representation for the location of the LALINET stations in South America. Argentina (AR): 1-) SMN Headquarters (Buenos Aires), 2-) CEILAP Headquarters (Buenos Aires), 3-) Comodoro Rivadavia (Chubut), 4-) Neuquén (Neuquén), 5-) Pilar (Cordoba), 6-) Río Gallegos airport (Santa Cruz), 7-) OAPA Río Gallegos (Santa Cruz), 8-) San Carlos de Bariloche (Río Negro), 9-) San Miguel de Tucumán (Tucumán). Bolivia (BO): 10-) La Paz (La Paz). Brazil (BR): 11-) Manaus (Amazonas), 12-) São Paulo (São Paulo), 13-) Cubatão (São Paulo), 14-) Natal (Rio Grande do Norte). Chile (CH): 15-) Punta Arenas (Magallanes), 16-) Temuco (Cautín). Colombia (CO): 17-) UNAL Medellín (Antioquia), 18-) SIATA Medellín (Antioquia), 19-) Cali (Valle del Cauca). Edited using Google my maps [14].

Developing a regional ground-based lidar network in Latin and South America is of strategic importance: The knowledge rendered by the high-resolution profiles allows the knowledge of a wide variety of atmospheric phenomena to complement satellite observations and other retrievals by diverse ground-based instruments. Unfortunately, the available infrastructure of lidar stations in Latin America is limited in certain aspects. For example, only a few stations operate regularly (contrasted to Europe and North America), stations have different instrument designs, radiosonde launchings are not occurring nearby all stations, and only a reduced number of sun photometers is distributed across the continent [7, 11]. To get around such limitations and consolidate standard protocols of measurements, data acquisition, quality control, and assurance routines, and data analysis, the Latin America Lidar Network, LALINET, was established in 2001, during the First Workshop on Lidar Measurements in Latin America, held in Camagüey, Cuba, in March 2001 [7, 11, 13]. It was recognized as being part of the GAW (Global Atmospheric Watch) Aerosol Lidar Observation Network (GALION) in 2013 [7, 11, 13]. **Figure 1** shows the location of the LALINET stations [14].

The next sections of this chapter will present information about mesospheric, stratospheric, and tropospheric monitoring by LALINET stations and teams around South America and Cuba, plus some significant results. **Table 1** below shows the operational stations and their characteristics. A detailed description of LALINET origin and its evolution is given in Ref. [7]. The Letter of Agreement between LALINET and GAW can be found in Ref. [15].

Country, City, Location Coordinates, Altitude (a.s.l.)	System configuration		
	Instrument	Emits (nm)	Detects (nm)
AR, Buenos Aires, SMN 34.5641 S, 58.4171 W, 10 m	Elastic Polarized	1064, 532, 355	1064, 532 (‖, ⊥), 355 (‖, ⊥)
AR, Buenos Aires, CEILAP 34.5553 S, 58.5062 W, 26 m	HSRL	1064, 532, 355	1064, 607, (HSRL, ‖, ⊥), 408, 387, 355 (‖, ⊥)
AR, Rivadavia, CRD Airport 45.7922 S, 67.4629 W, 48 m	Elastic Polarized	1064, 532, 355	1064, 532 (‖, ⊥), 355 (‖, ⊥)
AR, Neuquén, NQN Airport 38.9521 S, 68.1368 W, 266 m	Elastic Polarized	1064, 532, 355	1064, 532 (‖, ⊥), 355 (‖, ⊥)
AR, Pilar, OMGP 31.6755 S, 63.8730 W, 332 m	HSRL	1064, 532, 355	1064, 607, 532 (HSRL, ‖, ⊥), 408, 387, 355 (‖, ⊥)
AR, R. Gallegos, RGL Airport 51.6117 S, 69.3072 W, 17 m	Elastic Polarized	1064, 532, 355	1064, 532 (‖, ⊥), 355 (‖, ⊥)
AR, Río Gallegos, OAPA 51.6004 S, 69.3194 W, 19 m	DIAL	355 (Nd:YAG), 308 (Xe:Cl)	387, 355, 347, 332, 308
AR, Bariloche, BRC Airport 41.1473 S, 71.1640 W, 837 m	Raman	1064, 532, 355	1064, 532, 408, 387, 355
AR, S. M. de Tuc., TMO 26.7871 S, 65.2068 W, 485 m	Elastic Polarized	1064, 532, 355	1064, 532 (‖, ⊥), 355 (‖, ⊥)
BO, La Paz, UMSA 16.5381 S, 68.0686 W, 3420 m	Scanning Elastic	532	532
BR, Manaus, Embrapa 2.8906 S, 59.9698 W, 80 m	Raman	355	408, 387
BR, São Paulo, IPEN 23.5607 S, 46.7398 W, 764 m	Raman	1064, 532, 355	1064, 532, 530, 408, 387, 355
BR, Cubatão, CEPEMA 23.8865 S, 46.4370 W, 8 m	Mobile Raman	532	532, 607

Country, City, Location Coordinates, Altitude (a.s.l.)	System configuration		
	Instrument	**Emits (nm)**	**Detects (nm)**
BR, Natal, UFRN 5.8431 S, 35.2043 W, 20 m	Elastic Polarized	1064, 532, 355	1064, 532 (‖, ⊥), 355
CH, Punta Arenas, UMAG 53.1344 S, 70.8802 W, 10 m	Raman Polarized	1064, 532, 355	1064, 607, 532 (‖, ⊥), 408, 387, 355 (‖, ⊥)
CH, Temuco, UFRO 38.7459 S, 72.6156 W, 108 m	Elastic	532	532
CO, Medellín, UNAL 6.2619 N, 75.5760 W, 1538 m	Elastic	1064, 532	1064, 532
CO, Medellín, SIATA 6.2017 N, 75.5784 W, 1502 m	Elastic Polarized	355	355 (‖, ⊥)
CO, Cali, CIBioFi-UniValle 3.3770 N 76.5337 W, 982 m	Elastic Polarized	1064, 532, 355	1064 (‖, ⊥), 532 (‖, ⊥), 355 (‖, ⊥)

Details about the contributing teams, measurement protocols, reports, and equipment can be found on the web page http://www.lalinet.org. Detection of polarized light in the parallel (‖) and perpendicular (⊥) directions are indicated.

Table 1.
LALINET operational stations and their characteristics [7–11].

3. Mesosphere

Meteors enter in the upper atmosphere at very high velocities (15–70 km s^{-1}), and the collisions with the atmospheric constituents cause flash heating until the particles melt and their chemical elements vaporize. This ablation process is responsible for the layers of metal atoms as Na, K, Fe, Mg, Ca, Si, among others, which occur globally in the mesosphere and lower thermosphere (MLT). This cosmic dust's primary sources are the sublimation of comets as they approach the Sun on their orbits through the solar system and the collisions between asteroids.

Lidar use for the upper stratosphere, mesosphere, and lower thermosphere investigations started in São José dos Campos, Brazil, in 1969 with a ruby laser operated at 694.3 nm. Clemesha and Rodrigues obtained the first aerosol profile using lidar in South America in 1971 [16]. The height range of measurement was 5 to 35 km due to the use of an 8 x 10″ receiver mirror. Later were obtained profiles up to 90 km in height using a 48″ mirror. In this work, high concentrations of aerosols were observed in the troposphere, a minimum just below the tropopause, around 15 km height, and higher concentrations in the lower stratosphere.

In 1972, when a new "handmade" dye laser became operational (see a Photo of this equipment in **Figure 2**), it was possible to start measurements of the Na layer in the MLT region, using Fabry-Perot interferometers and tuning the laser in the Na D2 line, 5890 Å, with a precision of 0.02 Å [17]. This system enabled the measurement of the mesospheric Na from 75 to 105 km of height [18]. The system continued to be operated regularly for long years obtaining the Na concentration at MLT region with different time and height resolutions, the stratospheric aerosol by Mie Scattering, and the atmospheric density and temperature from 30 to 65 km by Rayleigh scattering. In April 1975, 6 months after the eruption of Volcán de Fuego in Guatemala, a massive increase in aerosol loads was observed in São José dos Campos, which remained in the atmosphere for almost two years [19].

Through Na profiles between 82 and 99 km obtained with the laser beam directed alternately in three positions in the sky, it was possible to estimate the wind's speed in the mesosphere [20, 21]. The velocities vary with height in an oscillatory manner,

Figure 2.
The handmade dye laser for Na probing (it operates from 1972 to 1992). See also in the picture Dr. Barclay R. Clemesha (in memoriam), the project's head.

with the amplitude increasing with height. These wave-like formations vary slowly with time and might be produced by propagating tides in the atmosphere. These structures' common feature is their downward motion with time, consistent with the upward propagation of gravity wave energy. The more extended periods of oscillations are attributed to tides [22, 23]. Lidar measurements of the stratospheric aerosols enabled the observation of the eruption of El Chichón in México, eight months after in São José dos Campos, Brazil [24]. The transport of aerosols of the Pinatubo eruption was much more rapid and could be seen just 45 days after the eruption [25].

Research involving Na has included the first detection of the so-called Sporadic Sodium Layers [26]. The events occurred more frequently through periods of more significant meteor showers, especially in August. It is common to have sporadic E layers coincident with Na enhancement, which suggests that enhanced layers are generated by the wind shear distortion of Na clouds originated from meteor ablation. A significant result was that the long-lived sporadic layers appear to have a different nature from the short-lived ones. The difference is manifested in the more extensive duration and broader thickness and how the events are correlated with sporadic E layers [27].

In 1992, analyses of the vertical distribution of atmospheric Na layer with lidar showed a long-term trend of the centroid height, which decayed by approximately 700 meters between 1972 and 1987 [28]. However, from 1972 to 2001, the trend was 93 meters per decade. This new result appears dramatically diminishes the possibility of long-term cooling of the upper atmosphere [29].

In 1997 a new technique was developed to measure the Doppler temperature of the atmospheric Na layer by using a two-element birefringent filter together with a 0.2 nm free spectral range Fabry-Perot interferometer to produce a linewidth of about 20 pm. It produced a multi-line signal of the laser, with the lines spacing precisely equal to the separation of D2a and D2b transition of Na. With this assembly, it was possible to obtain the mesosphere's temperature with a 5 K precision, a height resolution of 1 km, and a time resolution of 6 minutes [30, 31]. Lately, in 2004 the lidar was equipped with a new laser, which permitted more precise measurements of the mesopause temperature (see the assembly in **Figure 3**) [32, 33]. Gravity wave's effects on the temperature in the mesopause were also studied [34, 35].

Figure 3.
Photo presenting the continuum narrowband tunable laser for Na concentration and Mesopause temperatures. It operated at São José dos Campos measuring mesopause temperature from 2007 to 2009 and Na concentration from 2006 to 2016. This photo was taken by Barclay R. Clemesha (in memoriam).

Several mesospheric dynamics studies involving other instruments like photometers, meteor radar, and onboard rocket instruments have been made [23, 36–39]. A mobile lidar has been developed to measure the Na concentration simultaneously with the volume emission profile for the NaD line of airglow in rocket campaigns in the Brazilian equatorial region of Alcântara (2.3728 S, 44.3965 W). An illustrative photo of this system is shown in **Figure 4**. This experiment allowed calculating the branching ratio of the reaction involved in the Na airglow [40].

Along the time, the São José dos Campos lidar underwent many modifications and upgrades. In 1993, the transmitter laser was upgraded with a commercial laser (see its illustration in **Figure 5**). With this upgrade, it was possible to use the Rayleigh signal from the clean atmosphere from 30 to 75 km (below the resonant Na signal) to measure the relative atmospheric density and the absolute temperature. These measurements have been used to study mesospheric temperature general behavior and the effects of atmospheric waves [41]. The long series of measurements have enabled long-term studies of the mesospheric Na, aerosols, and temperatures associated with global change [29, 42, 43]. A dual-beam Na/K lidar was assembled in São José dos Campos, Brazil, to extend the Na layer studies and improve the knowledge about metal layers in the MLT region. This system was installed owing to a cooperative agreement between the National Space Science Center (China) and the National Institute for Space Research (Brazil) in November of 2016.

The lidar uses two laser beams of 589 nm and 770 nm to simultaneously measure Na and K concentrations by the resonant scattering at MLT. The signal-to-noise ratio response allows 3 min time resolution and 96 m of height resolution in the profiles [44]. **Figure 6** shows the Na/K lidar during operation.

It is essential to point out that, up to the present time, this is the unique K lidar system operating in the Southern Hemisphere (SH). For the first time, it was presented the nocturnal and seasonal behavior of K and Na concentrations measured simultaneously at SH [44]. The seasonal variation of these two metals was determined, and it is interesting to note their different behavior even though both are alkali metals and come from meteor ablation. Semiannual variation is observed in both metal concentrations with different maxima: K shows its maxima around the

Figure 4.
Photo illustrating the INPE mobile lidar used during rocket campaigns in the Brazilian equatorial region of Alcântara, on 31 may 1992.

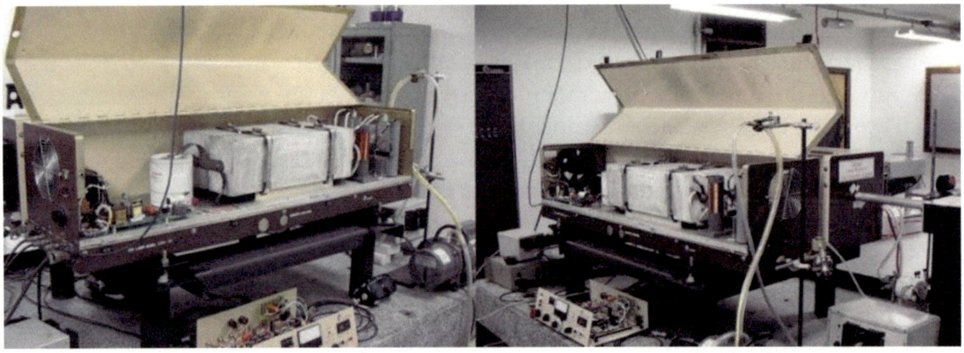

Figure 5.
Photos showing the candela laser system assembled at INPE São José dos Campos in 1993. This system operated between 1993 and 2006—Photos taken by B. R. Clemesha (in memoriam).

solstices more pronounced around June, and Na concentration shows a maximum around May and a broad one centered in September [44]. A plausible interpretation of the different seasonal changes between Na and K concentrations is presented in Ref. [45]. This analysis is based on two points: 1) the neutralization of $K+$ ions is particularly favored at low temperatures through summer (North Hemisphere), and 2) cycling between K and its primary neutral reservoir KHCO3 is substantially temperature independent [44]. Unfortunately, the first argument is not significant for this latitude, where the mesopause temperature has not a great summer to winter variation [33].

Figure 6.
Picture showing the dual-beam Na/K lidar located at São José dos Campos, Brazil. The vertical orange beam is at 589 nm for Na scattering and the infrared one at 770 nm for K scattering. This last is not visible in the photo, but the red star indicates the beam position. Liu Zhengkuan took the original photo.

4. Stratosphere

4.1 Historical overview

The first lidar measurements concerning stratospheric aerosols in Latin America were performed in Kingston, Jamaica, between 1964 and 1979 [46]. The lidar system held for these measurements was managed by the University of the West Indies and supported by the US Air Force [47]. Its primary purpose was to investigate the atmospheric profile, measuring molecular scattering. Moreover, the system proved valuable for measurements of stratospheric aerosol layers at wavelength 694 nm [48]. These lidar measurements from Jamaica represented a pioneering role, concomitantly with different research teams, developing lidars' capacities to measure aerosols in the lower stratosphere [49]. Those measurements were also an essential contribution to the stratosphere's early studies in the tropics [50].

In 1969, a new lidar instrument was designed and developed at INPE by Prof. Barclay Clemesha (see Section 3 for details). This equipment's primary objective was to investigate the mesosphere dynamics; besides, stratospheric aerosol measurements were also performed. The first measurements were carried in 1970 at wavelength 694 nm [16], and regular measurements began in 1972 [51]. This project was responsible for collecting the longest stratospheric aerosol profile measurements in Latin America and the Southern Hemisphere's tropical zone, extending from 1971 to 2016. It includes stratospheric aerosols profiles from the two more significant volcanic eruptions of the XX century second half: the first happened in Mexico on 04 April 1982 (El Chichón), and the second in the Philippines on 14 June 1991 (Mount Pinatubo) [51, 52]. Measurements conducted at INPE between 1972 and 2016 proved the value and the importance of the stratospheric aerosols' long-term monitoring. They have rendered information to understand the stratospheric aerosols layer evolution in the Southern Hemisphere's tropics since the '50s [53].

A Cuban-Soviet scientific cooperation agreement supported the deployment in 1988 of a lidar system designed for stratospheric aerosols measurements at the Camagüey Meteorologic Center in Cuba [54]. The instrument operated intermittently between 1988 and 1997, providing stratospheric aerosols measurements from the Mount Pinatubo eruption in 1991. The 1988–1990 lidar aerosol profiles, at 532 nm, combined with satellite measurements, have been used to study background stratospheric aerosols in the Caribbean [55]. Camagüey Lidar Station (CLS) stratospheric aerosols profiles from Mount Pinatubo also contributed to the study of the radiative impacts of the eruption at regional [56] and global [57] scales. Moreover, the Camagüey lidar database was also used to validate the stratospheric aerosol SAGE II satellite measurements from Mount Pinatubo eruption [58, 59]. Furthermore, it was used to generate an extinction climatology in the UV for correcting Brewer ozone measurements [60].

By 1994 the Laser and Applications Research Center (CEILAP - UNIDEF) in Buenos Aires, Argentina, developed various lidar systems for atmospheric research [7]. One of these devices was designed to measure the atmospheric boundary layer, cirrus clouds, and tropospheric aerosols, operating at wavelength 532 nm [61]. A collaborative study between CEILAP and CLS evaluated how this lidar system could also be used for the higher troposphere and lower stratospheric aerosols research. Upon analyzing two tropospheric aerosols profiles extending into the lower stratosphere, encouraging results were found [62]. In June 2005, another lidar system was designed and installed by CEILAP in Río Gallegos, Patagonia. This instrument's primary goal was performing measurements of stratospheric ozone, tropospheric and stratospheric aerosols, and water vapor. In particular, stratospheric aerosol profiles are used to correct the stratospheric ozone [63].

Western South America is bordered by the Andes, which divides the continent into two distinct regions. In South America, the vast majority of active volcanoes are located in the eastern part of the continent, and ash eruptions are routinely reported throughout the region. The volcanic activity includes periods of ash eruptions and cycling eruptions that spread out over months or even years [64, 65]. Great active volcanoes in South America are Nevado del Ruíz, in Colombia; Cotopaxi, Tungurahua, and Reventador, in Ecuador; Villarrica, Llaima, Nilahue, Lascar, Chaitén, and Calbuco, in Chile; El Misti, Ubinas and Sabancaya, in Peru; Aracar, Copahue, and Planchón-Peteroa in Argentina. There are no reported active volcanoes in Paraguay, Uruguay, Venezuela, Guyana, Suriname, and Brazil [64, 65].

On 22 April 2015, in Chile, the Calbuco volcano erupted and injected a significant amount of ashes and aerosols into the atmosphere [66].

The volcanic aerosol profiles in both the upper troposphere and the lower stratosphere, which originated from the Calbuco volcano eruption in Chile on 22 April 2015, were measured by different lidar stations in South America [7]. It was the first time that LALINET lidar stations, distributed across the continent, could analyze aerosol profiles together during an event. Lidar stations located in Buenos Aires, Comodoro Rivadavia, San Carlos de Bariloche, Neuquén, and Rio Gallegos (all five in Argentina), Concepción (Chile), and São Paulo (Brazil) observed the aerosols profiles [7, 67]. LALINET stations' capabilities to operate in a coordinated way in case of a volcanic eruption were challenged, highlighting the coordination among LALINET teams.

On 23 April 2015 (one day after the eruption), the lidar system at the University of Concepción measured the aerosols profiles between 5 and 9 km, showing a multilayer structure. Both layers merged at around 7 km, decreasing its intensity and narrowing. The following day 24 April 2015, the two layers registered in the day before at Concepción were detected in the nighttime by the lidar system placed in Buenos Aires, Argentina, in heights varying between 5 and 7 km showing a drowning leaning. The aerosol's multilayer formation was present at both lidar sites when identified for the first time. Lidar measurements conducted at IPEN in São Paulo on 27 April 2015 (five days after the eruption) exhibited aerosols found at an altitude of about 19 km in the stratosphere (**Figure 7**) [66]. Those lidar extinction profiles were confronted with those measured by the Ozone Mapping and Profiler Suite Limb Profiler (OMPS/LP) instrument, revealing promising results [7].

4.2 Differential absorption lidar measurements in Argentina

The behavior of trace constituents in the Earth's upper atmosphere, dictated by diverse physical processes, is of particular interest for the balance of stratosphere and mesosphere. Expressly, ozone has a principal function by absorbing the short-wavelength UV radiation (which might damage life) and keep the radiative budget stable [68]. For those reasons, ozone has been at the focus of the middle atmosphere research effort [69, 70].

Researchers' interest in performing lidar measurements from the southern region of the southern hemisphere dates back to 1995. Researchers from CEILAP, together with Prof. Gérard Mégie (who was then head of the *Service d'Aéronomie* in France), considered conducting a campaign to measure ozone profiles using a DIAL (differential absorption lidar) system, in Patagonia, Argentina [71]. The configuration and installation of the lidar system began as a collaboration linking the two institutions. For the DIAL technique, two laser wavelengths are used to measure atmospheric ozone. One wavelength is well absorbed by ozone, while the other not. After the wavelengths travel into the atmosphere and are backscattered

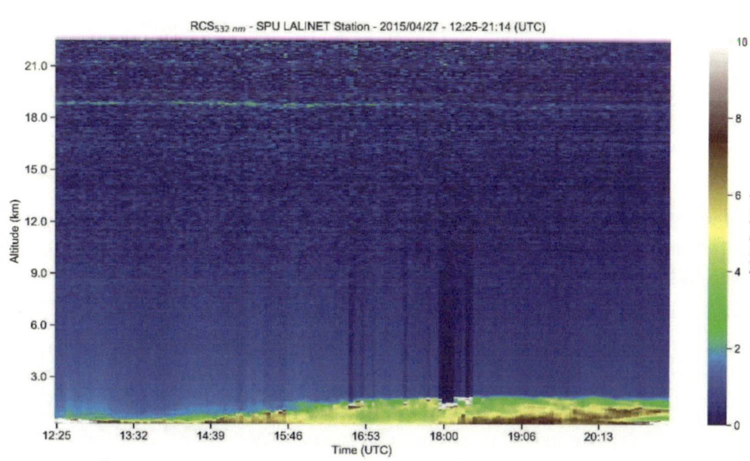

Figure 7.
Quick-look of the RCS at 532 nm measured at SPU Lidar Station on 27 April 2015. The SPU Lidar Station is installed at the Center for Lasers and Applications of the nuclear and energy research institute (CELAP/IPEN) in São Paulo. The signal between 18 km and 20 km shows aerosols originating from the Calbuco volcano eruption on 22 April 2015, in Chile.

to two receivers, it is possible to make a ratio of the measurements, allowing direct determination of the ozone concentration as a function of range.

The instrument became operational in 1997 in Villa Martelli, Buenos Aires, where the headquarters of CEILAP is located. The initial version had only one telescope, which was 50 cm in diameter. It operated successfully until 2002. Later, the number of telescopes was increased to four, and a spectrometer was added. The apparatus was fine-tuned at the Villa Martelli headquarters.

The *Service d'Aéronomie* loaned the equipment's electronic project and a container, which had already been used in the Arctic. However, financing was still an issue. Fortunately, since 1999, CEILAP has cooperated with the Tohoku Institute of Technology in Sendai, Japan. The Japan International Cooperation Agency (JICA) supported the south's entire measurement campaign. It further contributed to acquiring a new Nd:YAG laser, which is imperative to the DIAL instrument. In this way, the SOLAR (stratospheric ozone lidar of Argentina) campaign started in June 2005 [72].

The campaign's feasibility study was conducted, considering the nocturnal cloud cover over four towns in Argentine Patagonia. The data were compared with those corresponding to days when the Antarctic polar vortex crosses over these towns.

Different tracers were also considered, such as the total ozone column values measured by total ozone mapping spectrometry, the equivalent latitude method, and the potential vorticity maps calculated for the mid-stratosphere, according to studies carried out in collaboration with the Service d'Aeronomie in France and the National Institute for Environmental Studies in Japan.

The city of Río Gallegos region met the necessary conditions for the measurements. It is located at 2612 km from Buenos Aires, on the River Gallegos estuary banks, and has 140,000 inhabitants. Like other cities in southern Argentina and Chile, Río Gallegos is reached by the ozone hole's edge during the austral spring. However, compared with its counterparts, it has a more significant number of clear nights or nights with less than one-eighth cloud cover, which means more opportunities for making measurements with the ozone DIAL. Río Gallegos also hosts the National University of Southern Patagonia, whose staff could participate in the campaign, and is near to Punta Arenas, Chile, where another research group has used a Brewer instrument to make ozone measurements, in cooperation with Brazilian researchers. On 10 June 2005, the team set off overland for Río Gallegos in two trucks that traveled 2612 km from Buenos Aires to the Military Air Base in Río Gallegos, where a mobile laboratory was set up. The base is located 18 km from the center of the town [72, 73].

A Xe:Cl excimer laser emission at 308 nm is employed for the absorbed wavelength in the DIAL technique, and an Nd:YAG laser at 355 nm third harmonic line is employed as the reference wavelength. Six channels are used for signal acquisition [72]. Four of them detect the emitted wavelengths' elastically backscattered signal (high energy mode for the higher altitude ranges, attenuated energy for the lower ranges), and two correspond to the Raman wavelengths [72]. The CEILAP's DIAL instrument setup is shown in **Figure 8**, and its full description can be found in Ref. [10].

The CEILAP Lidar Division, in cooperation with other national and international institutions, has organized the SOLAR (Stratospheric Ozone Lidar of ARgentina) Campaign as a part of environmental investigations in the Southern Hemisphere [72]. This campaign's objective was to monitor different atmospheric constituents using remote sensing techniques, mainly related to lidar, in Argentina's southern part. The most critical and complex instrument involved in this campaign is a differential absorption lidar capable of producing precise and accurate stratospheric ozone profiles [72, 73].

The most substantial decrease of the ozone column over Río Gallegos through the 2005 spring was observed on 8 October, with a total ozone column of 196 DU

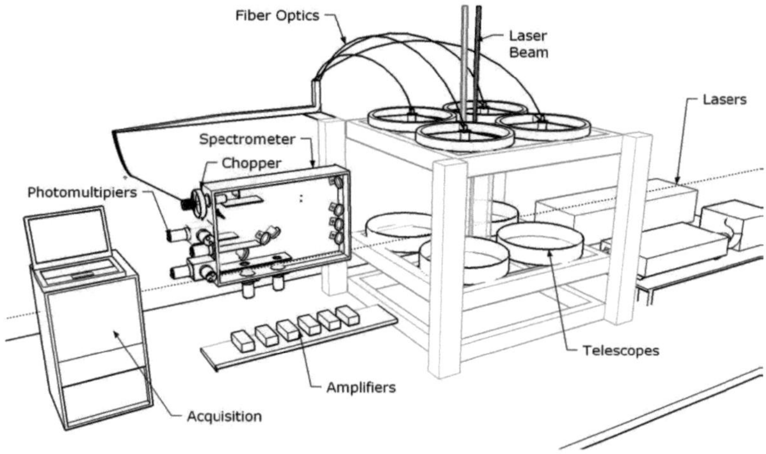

Figure 8.
Experimental setup of differential absorption lidar (DIAL) developed at CEILAP.

estimated from integrating an ozone profile based on the lidar measurement and the US Standard 1976. This value expresses a decrease of 45% in the total ozone column concerning the mean total ozone value outside the ozone hole for this month (about 350 DU). **Figure 9** shows the measured lidar profile on this day (dashed line), together with the ozone profile measured on 17 October (dotted line), which corresponds to standard ozone conditions outside the ozone hole (about 357 DU). The figure also shows the climatologic profile (black line) from the SAGE II measurements, which corresponds to the mean of the ozone measurements outside the ozone hole for the 1995–2004 period.

From the full set of lidar measurements, were selected 37 lidar profiles that match the HRLS profiles. The monthly mean lidar profiles were confronted with similar profiles measured by the High-Resolution Dynamics Limb Sounder (HIRDLS) device onboard the NASA-Aura satellite. The collocation criteria for selecting satellite data were set using a distance of up to 500 km from site measurement and a temporal selection of about twelve hours for the measurement time.

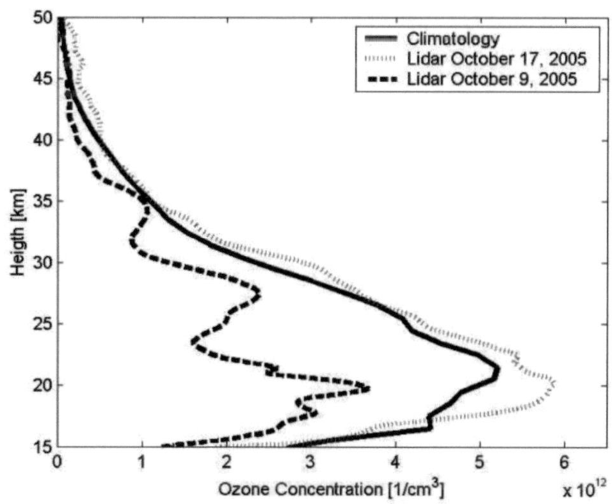

Figure 9.
Lidar ozone profile inside (dashed line) and outside (gray dotted line) ozone hole in Río Gallegos. Climatologic profile for October from SAGE II data (black line) [74].

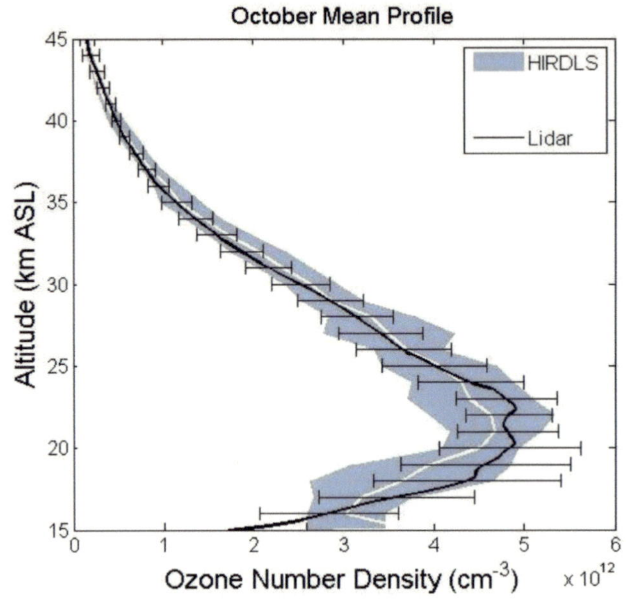

Figure 10.
Mean lidar profile (black line - error bar corresponds to ±1 std) and mean HIRDLS (white line) ±1 std. (shadow area) for October.

The mean stratospheric ozone lidar profile for October in Río Gallegos is shown in **Figure 10**. For comparison, the same quantity from satellite data is included.

In general, good agreement between lidar and satellite data was found (inside the statistical error bar, with a relative difference of around 10%). The maximum disagreement between lidar and satellite data was observed in August mean profiles around 30 km. For October, the agreement was better than 10% above the ozone peak concentration. In general, it was observed that the variability of lidar profile concentrations is higher around the ozone peak, decreasing with height.

Differential Absorption lidar techniques have been demonstrated to be a reliable remote sensing technique to retrieve the stratosphere's ozone profile [73]. Argentina has used DIAL techniques since 1999. In 2005, with French and Japanese researchers' collaboration, the Lidar Division of CEILAP established a new site in Southern Patagonia, the South Patagonia Atmospheric Observatory (OAPA). This device has been part of Network Data for Atmospheric Composition Change (NDACC) since 2008, and the research using its measurements allows the study of ozone hole overpass from South America [75] and the satellite validation in the South Hemisphere. After the SOLAR Campaign, several initiatives were carried out related to stratospheric ozone monitoring in Argentina. For example, the UVO3-Patagonia (2008–2010) and SAVER-Net projects (2013–2018) were the research activities made in collaboration with JICA, and Japanese and Chilean Researchers went more in-depth the observation of ozone in vertical profiles and total ozone column [76].

5. Conclusions

Part I of this chapter offered the opportunity to give a scientific overview of current and past lidar observation activities conducted in South America, with Cuba's participation. This overview spans over almost 50 years of activities and grants how this part of the world is concerned with laser remote sensing of the atmosphere in almost its whole structure: Mesosphere and Stratosphere. This top-down approach also followed a chronological delivery of results, with the first results coming from the region in the highest portion of the atmosphere (mesosphere), and going downwards to stratospheric, and finally at the tropospheric studies. If, in the first years, these activities started as individual initiatives at different countries and research groups levels, the creation of a federative lidar network, namely LALINET, helped somehow to have more coordinated measurements. Moreover, the implementation of SAVERNET in Argentina and Chile improved how these joint measurements are conducted. The studies conducted in the mesosphere account for one of the most extended time series of lidar data, being of great importance in the Southern Hemisphere. Also, significant results about Na and K concentrations and their variability over almost three decades are available. The studies of ozone concentration in the stratosphere also provided relevant results, unprecedented for this portion of the globe. Part II of this chapter will be dedicated to tropospheric lidar observations.

Acknowledgements

The authors are thankful to the Brazilian Agencies National Council for Scientific and Technological Development (CNPq), Coordination for the Improvement of Higher Education Personnel (CAPES), São Paulo Research Foundation (FAPESP), Brazilian Agricultural Research Corporation (EMBRAPA), and National Institute of Amazonian Research (INPA) LBA Central Office in Manaus. The authors also thank the NASA/AERONET teams, Japan International Cooperation Agency (JICA), the Argentine Agencies National Scientific and Technical Research Council (CONICET), National Agency for the Promotion of Research, Technological Development and Innovation (ANPCyT), the Argentine National Defense University (UNDEF), UNDEFI and PID-UTN Projects, the Ministry of Defense of Argentina, and the French National Centre for Scientific Research (CNRS). Also, to all NASA's technical personnel, the Argentine Institute of Scientific and Technical Research for Defense (CITEDEF), and the Argentine National Meteorological Service (SMN), who have kept the solar photometers in operation, and especially to Raúl D'Elia. The authors wish to acknowledge the entire NASA CALIPSO and MODIS (AQUA/TERRA) teams, the NOAA Air Resources Laboratory, for providing the HYSPLIT transport and dispersion model and the READY website, ESA/EOM projects teams, the Suomi NPP (National Polar-orbiting Partnership) Mission teams, and the Sentinel 5-P TROPOMI team. The authors also acknowledge the financial support from CIBioFi, the Colombian Science, Technology, and Innovation Fund-General Royalties System (Fondo CTeI-Sistema General de Regalías), and Gobernación del Valle del Cauca. The authors acknowledge the China-Brazil Joint Laboratory for Space Weather (CBJLSW) for Supporting this Book Chapter. Vania F. Andrioli would like to thank the CBJLSW and the National Space Science Center (NSSC) of the Chinese Academy of Sciences (CAS) for supporting her postdoctoral fellowship. The authors from the Universidad de Magallanes would like to acknowledge the financial support of the Japan Science and Technology Agency (JST) / Japan International Cooperation Agency (JICA), the Science and Technology Research Association for Sustainable Development (SATREPS) through the SAVERNet project; and the Program FONDECYT of the Chilean National Agency for Research and Development (ANID) through Project FONDECYT 11181335.

Conflict of interest

The authors declare no conflict of interest.

Author details

Eduardo Landulfo[1*], Alexandre Cacheffo[1,2], Alexandre Calzavara Yoshida[1,2], Antonio Arleques Gomes[1], Fábio Juliano da Silva Lopes[1,3], Gregori de Arruda Moreira[1,4,5], Jonatan João da Silva[1,6], Vania Andrioli[7,8], Alexandre Pimenta[7], Chi Wang[9], Jiyao Xu[9], Maria Paulete Pereira Martins[7], Paulo Batista[7], Henrique de Melo Jorge Barbosa[10,11], Diego Alves Gouveia[10,12], Boris Barja González[13], Felix Zamorano[13], Eduardo Quel[14], Clodomyra Pereira[15,16], Elian Wolfram[17,18], Facundo Ismael Casasola[15,16,19], Facundo Orte[18], Jacobo Omar Salvador[18], Juan Vicente Pallotta[18], Lidia Ana Otero[14,19], Maria Prieto[15,16], Pablo Roberto Ristori[14], Silvina Brusca[14], John Henry Reina Estupiñan[20,21], Estiven Sanchez Barrera[20], Juan Carlos Antuña-Marrero[22], Ricardo Forno[23], Marcos Andrade[23], Judith Johanna Hoelzemann[24], Anderson Guimarães Guedes[25], Cristina Tobler Sousa[24], Daniel Camilo Fortunato dos Santos Oliveira[26,27], Ediclê de Souza Fernandes Duarte[27], Marcos Paulo Araújo da Silva[26,27] and Renata Sammara da Silva Santos[1,24]

1 Center for Lasers and Applications (CELAP), Institute of Energy and Nuclear Research (IPEN), Sao Paolo, Brazil

2 Institute of Exact and Natural Sciences of Pontal (ICENP), Federal University of Uberlândia (UFU), Ituiutaba, Brazil

3 Institute of Environmental, Chemical and Pharmaceutical Sciences (ICAQF), Federal University of São Paulo (UNIFESP), Diadema, Brazil

4 Federal Institute of São Paulo (IFSP), Campus Registro, Sao Paolo, Brazil

5 Institute of Astronomy, Geophysics and Atmospheric Sciences (IAG), University of São Paulo (USP), Sao Paolo, Brazil

6 Center for Exact Sciences and Technologies (CCET), Federal University of Western Bahia (UFOB), Barreiras, Brazil

7 National Institute for Space Research (INPE), São José dos Campos, Brazil

8 China-Brazil Joint Laboratory for Space Weather (NSSC/INPE), São José dos Campos, Brazil

9 State Key Laboratory of Space Weather (SKSW), National Space Science Center (NSSC), Chinese Academy of Sciences (CAS), Beijing, China

10 Physics Institute, University of São Paulo (USP), Sao Paolo, Brazil

11 Physics Department, University of Maryland Baltimore County (UMBC), Baltimore, USA

12 Royal Netherlands Meteorological Institute (KNMI), De Bilt, The Netherlands

13 Department of Mathematics and Physics, University of Magallanes (UMAG), Punta Arenas, Chile

14 Institute of Scientific and Technical Research for Defense (CITEDEF) - UNIDEF (MINDEF – CONICET), Buenos Aires, Argentina

15 Military Geographical Service, National Geographic Institute (IGN), Buenos Aires, Argentina

16 General Directorate of Research and Development of the Argentine Army (DIGID), Buenos Aires, Argentina

17 Argentine National Weather Service (SMN), Buenos Aires, Argentina

18 Laser and Applications Research Center (CEILAP), UNIDEF (MINDEF–CONICET), Buenos Aires, Argentina

19 Argentine National Defense University (UNDEF), Army Engineering Faculty (FIE), Buenos Aires, Argentina

20 Centre for Bioinformatics and Photonics (CIBioFi), Universidad del Valle (UniValle), Cali, Colombia

21 Physics Department, Universidad del Valle (UniValle), Cali, Colombia

22 Department of Theoretical, Atomic and Optical Physics, University of Valladolid (UVA), Valladolid, Spain

23 Department of Physics, Major University of San Andrés (UMSA), La Paz, Bolivia

24 Department of Atmospheric and Climate Sciences (DCAC), Federal University of Rio Grande do Norte (UFRN), Natal, Brazil

25 School of Science and Technology (ECT), Federal University of Rio Grande do Norte (UFRN), Natal, Brazil

26 Polytechnic University of Catalonia (UPC), Barcelona, Spain

27 Graduate Program in Climate Sciences, Federal University of Rio Grande do Norte (UFRN), Natal, Brazil

*Address all correspondence to: landulfo@gmail.com; elandulf@ipen.br

References

[1] The Intergovernmental Panel on Climate Change. About. Working Groups and Reports. Available from: http://www.ipcc.ch/. Accessed 22 September 2020.

[2] The Nobel Peace Prize, 2020: The Nobel Peace Prize Winners of 2007. Available from: https://www.nobelpeaceprize.org/Prize-winners/Winners/2007. Accessed 22 September 2020.

[3] Haywood, J., Boucher, O., 2000: Estimates of the direct and indirect radiative forcing due to tropospheric aerosols: A review. Reviews of Geophysics, v. 38, Issue 4, 513-543. https://doi.org/10.1029/1999RG000078.

[4] Lohmann, U., Feichter, J., 2005: Global indirect aerosol effects: A review. Atmospheric Chemistry and Physics, v. 5, Issue 3, 715-737. https://doi.org/10.5194/acp-5-715-2005.

[5] Kovalev, W. A., Eichinger, W. E., 2004: Elastic Lidar: Theory, Practice, and Analysis Methods. New Jersey: Wiley-Interscience Publication. John Wiley & Sons, Inc. ISBN: 9780471201717. https://doi.org/10.1002/0471643173.

[6] Weitkamp, C., editor. 2005: Lidar: Range-Resolved Optical Remote Sensing of the Atmosphere. Springer Series in Optical Sciences. New York: Springer-Verlag. ISBN: 978-0-387-40075-4. https://doi.org/10.1007/b106786.

[7] Antuña, J. C., Landulfo, E., Estevan, R., Barja, B., Robock, A., Wolfram, E., Ristori, P., et al., 2017: LALINET: The First Latin American–Born Regional Atmospheric Observational Network. Bull. Amer. Meteor. Soc., v. 98, Issue 6, 1255-1275. https://doi.org/10.1175/BAMS-D-15-00228.1.

[8] LALINET, 2014: Report of the inversion unified algorithm working group: I Workshop on Lidar Inversion Algorithms-LALINET. Available from: http://lalinet.org/uploads/Analysis/Concepcion2014/report_1_analysis_workshop.pdf. Accessed 22 September 2020.

[9] Landulfo, E., Lopes, F. J. S., Moreira, G. A., Marques, M. T. A., Osneide, M., Antuña, J. C., et al., 2016: ALINE/LALINET network status. EPJ Web of Conferences, v. 119, 19004. Proc. 27th International Laser Radar Conference (ILRC 27). https://doi.org/10.1051/epjconf/201611919004.

[10] Wolfram, E. A., Salvador, J., D'Elia, R., Casiccia, C., Paes Leme, N., Pazmiño, A., et al., 2008: New differential absorption lidar for stratospheric ozone monitoring in Patagonia, South Argentina. Journal of Optics A: Pure and App. Optics, v. 10, Issue 10, 104021. https://doi.org/10.1088/1464-4258/10/10/104021.

[11] Guerrero-Rascado, J. L., Landulfo, E., Antuña, J. C., Barbosa, H. M. J., Barja, B., Bastidas, A. E., et al., 2016: Latin American Lidar Network (LALINET) for aerosol research diagnosis on network instrumentation. J. Atmos. Sol.-Terr. Phys., v. 138-139, 112-120. https://doi.org/10.1016/j.jastp.2016.01.001.

[12] UNESCO, 2020: World Heritage Centre. Latin America and the Caribbean. Available from: http://whc.unesco.org/en/lac/. Accessed 29 September 2020.

[13] Rodriguez, A. R., Antuña, J. C., 2017: Standardizing the determination of the molecular backscatter coefficient profiles for LALINET lidar stations using ERA-Interim Reanalysis. Óptica Pura y Aplicada, v. 50, Issue 1, 103-113. https://doi.org/10.7149/OPA.50.1.49013.

[14] Google My Maps, 2020. Map data ©2020 INEGI Imagery ©2020 NASA,

TerraMetrics. South America and LALINET Stations, 1:1000. Edited. Available from: http://lalinet.org. Accessed 6 November 2020.

[15] LALINET. Letter of Agreement, 2013: Available from: LALINET website: http://www.lalinet.org/index.php/Aline/Commitment. Accessed 29 September 2020.

[16] Clemesha, B. R., Rodrigues, S. N., 1971: The stratospheric scattering profile at 23° South. Journal of Atmospheric and Terrestrial Physics, v. 33, Issue 7, 1119-1124. https://doi.org/10.1016/0021-9169(71)90132-2.

[17] Clemesha, B. R., Kirchhoff, V. W. J. H., Simonich, D. M., 1975: Automatic wavelength control of a flashlamp-pumped dye laser. Optical and Quantum Electronics, v. 7, 193-196. https://doi.org/10.1007/BF00619592.

[18] Kirchhoff, V. W. J. H., Clemesha, B. R., 1973: Atmospheric Sodium Measurements at 23° S. Journal of Atmospheric and Terrestrial Physics, v. 35, Issue 8, 1493-1498. https://doi.org/10.1016/0021-9169(73)90150-5.

[19] Clemesha, B. R., Simonich, D. M., 1978: Stratospheric Dust Measurements, 1970-1977. Journal of Geophysical Research, v. 83, Issue C5, 2403-2408. https://doi.org/10.1029/JC083iC05p02403.

[20] Clemesha, B. R., Kirchhoff, V. W. J. H., Simonich, D. M., Batista, P. P., 1981: Mesospheric winds from lidar observations of atmospheric sodium. Journal of Geophysical Research, v. 86, Issue A2, 868-870. https://doi.org/10.1029/JA086iA02p00868.

[21] Clemesha, B. R., Kirchhoff, V. W. J. H., Simonich, D. M., 1981: Remote measurement of tropospheric and stratospheric winds by ground-based lidar. Applied Optics, v. 20, Issue 17, 2907-2910. https://doi.org/10.1364/AO.20.002907.

[22] Clemesha, B. R., Kirchhoff, V. W. J. H., Simonich, D. M., Takahashi, H., Batista, P. P., 1980: Spaced lidar and nightglow observations of an atmospheric sodium enhancement. Journal of Geophysical Research: Space Physics, v. 85, Issue A7, 3480-3484. https://doi.org/10.1029/JA085iA07p03480.

[23] Batista, P. P., Clemesha, B. R., Kirchhoff, V. W. J. H., 1985: Tidal oscillations in the atmospheric sodium layer. Journal of Geophysical Research, v. 90, Issue D2, 3881-3888. https://doi.org/10.1029/JD090iD02p03881.

[24] Clemesha, B. R., Simonich, D. M., 1983: Lidar observations of the El Chichón dust cloud at 23° S. Geophysical Research Letters, v. 10, Issue 4, 321-324. https://doi.org/10.1029/GL010i004p00321.

[25] Simonich, D. M., Clemesha, B. R., 1995: Comparison between the El Chichón and Pinatubo aerosol clouds at 23° S. In: XXI IUGG General Assembly, Boulder, Colorado, EUA. http://www.iugg.org/assemblies/1995boulder/.

[26] Clemesha, B. R., Kirchhoff, V. W. J. H., Simonich, D. M., Takahashi, H., 1978: Evidence of an extra-terrestrial source for the mesospheric sodium layer. Geophysical Research Letters, v. 5, Issue 10, 873-876. https://doi.org/10.1029/gl005i010p00873.

[27] Batista, P. P., Clemesha, B. R., Batista, I. S., Simonich, D. M., 1989: Characteristics of the sporadic sodium layers observed at 23 S. Journal of Geophysical Research, v. 94, Issue A11, 15349-15358. https://doi.org/10.1029/JA094iA11p15349.

[28] Clemesha, B. R., Simonich, D. M., Batista, P. P., 1992: A long-term trend in the height of the atmosphere sodium layer possible evidence for global change. Geophysical Research Letters, v. 19, Issue 5, 457-460. https://doi.org/10.1029/92GL00123.

[29] Clemesha, B. R., Batista, P. P., Simonich, D. M., 2003: Long-term variations in the centroid height of the atmospheric sodium layer. Advances in Space Research, Oxford, v. 32, Issue 9, 1707-1711. https://doi.org/10.1016/S0273-1177(03)90466-2.

[30] Clemesha, B. R., Jorge, M. P. P. M., Simonich, D. M., Batista, P. P., 1997: A new method for measuring the Doppler temperature of the atmospheric sodium layer. Advances in Space Research, v. 19, Issue 4, 681-684. https://doi.org/10.1016/S0273-1177(97)00163-4.

[31] Clemesha, B. R., Veselovskii, I., Batista, P. P., Jorge, M. P. P. M., 1999: First mesopause temperature profiles from a fixed southern hemisphere site. Geophysical Research Letters, v. 26, Issue 12, 1681-1684. https://doi.org/10.1029/1999GL900342.

[32] Clemesha, B. R., Simonich, D. M., Batista, P. P., 2010: Mesopause region temperature structure observed by sodium resonance lidar. Journal of Atmospheric and Solar-Terrestrial Physics, v. 72, Issues 9-10, 740-744. https://doi.org/10.1016/j.jastp.2010.03.017.

[33] Clemesha, B. R., Simonich, D. M. and Batista, P. P., 2011: Sodium lidar measurements of mesopause region temperatures at 23° S. Advances in space research, v. 47, Issue 7, 1165-1171. https://doi.org/10.1016/j.asr.2010.11.030.

[34] Yang, G., Clemesha, B., Batista, P., Simonich, D., 2010: Seasonal variations of gravity wave activity and spectra derived from sodium temperature lidar. Journal of Geophysical Research, v. 115, Issue D18, D18104. http://doi.org/10.1029/2009JD012367.

[35] Andrioli, V. F., Batista, P. P., Xu, J., Yang, G., Chi, W., Zhengkuan, L., 2017: Strong temperature gradients and vertical wind shear on MLT region associated to instability source at 23° S. Journal of Geophysical Research: Space Physics, v. 122, Issue 4, 4500-4511. https://doi.org/10.1002/2016JA023638.

[36] Clemesha, B. R., Kirchhoff, V. W. J. H., Simonich, D. M., Takahashi, H., Batista, P. P., 1979: Simultaneous observations of sodium density and the NaD, OH (8, 3), and OI 5577-Å, nightglow emissions. Journal of Geophysical Research, v. 84, Issue A11, 6477-6482. https://doi.org/10.1029/JA084iA11p06477.

[37] Clemesha, B. R., Batista, P. P., Batista, I. S., 1998: Lidar observations of atmospheric sodium at an equatorial location. Journal of Atmospheric and Solar-Terrestrial Physics, v. 60, Issue 18, 1773-1778. https://doi.org/10.1016/S1364-6826(98)00144-8.

[38] Clemesha, B. R., Batista, P. P., 2001: Simultaneous measurements of meteor winds and sporadic sodium layers in the 80-110 km region. Advances in Space Research, v. 27, Issue 10, 1679-1684. https://doi.org/10.1016/S0273-1177(01)00238-1.

[39] Clemesha, B. R., Batista, P. P., Simonich, D.M., 2002: Tide-induced oscillations in the atmospheric sodium layer. Journal of Atmospheric and Solar-Terrestrial Physics, v. 64, Issues 12-14, 1321-1325. https://doi.org/10.1016/S1364-6826(02)00115-3.

[40] Clemesha, B. R., Takahashi, H., Melo, S. M. L., 1993: A simultaneous measurement of the vertical profiles of sodium nightglow and atomic sodium density in the upper atmosphere. Geophysical Research Letters, v. 20, Issue 13, 1347-1350. https://doi.org/10.1029/93gl01121.

[41] Batista, P. P., Clemesha, B. R., Simonich, D. M., 2008: Tidal associated temperature disturbances observed in the middle atmosphere (30-65 km) by a Rayleigh Lidar at 23° S. Advances

in Space Research, v. 41, Issue 9, 1408-1414. https://doi.org/10.1016/j.asr.2006.12.002.

[42] Deshler, T., Anderson-Sprecher, R., Jäger, H., Barnes, J., Hoffman, D. J., Clemesha, B. R., et al., 2006: Trends in the non-volcanic component of stratospheric aerosol over the period 1971-2004. Journal of Geophysical Research, v. 111, Issue D1, D01201. https://doi.org/10.1029/2005JD006089.

[43] Batista, P. P., Simonich, D. M., Clemesha, B. R., 2009: A 14-year monthly climatology and trend in the 35-65 km altitude range from Rayleigh Lidar temperature measurements at a low latitude station. Journal of Atmospheric and Solar-Terrestrial Physics, v. 71, Issue 13, 1456-1462. https://doi.org/10.1016/j.jastp.2008.03.005.

[44] Andrioli, V. F., Xu, J., Batista, P. P., Pimenta, A. A., Resende, L. C. A., Savio, et al., 2020: Nocturnal and seasonal variation of Na and K layers simultaneously observed in the MLT Region at 23°S. Journal of Geophysical Research: Space Physics, v. 125, Issue 3, e2019JA027164. https://doi.org/10.1029/2019JA027164.

[45] Plane, J. M. C., Feng, W., Dawkins, E., Chipperfield, M. P., Hoffner, J., Janches, D., Marsh, D. R., 2014: Resolving the strange behavior of extra-terrestrial potassium in the upper atmosphere. Geophysical Research Letters, v. 41, Issue 13, 4753-4760. https://doi.org/10.1002/2014GL060334.

[46] Philip, M. T., Alleyne, H., 1985: Lidar observations of the stratospheric aerosol layer over Kingston, Jamaica. J. Atmos. Sci., v. 42, Issue 9, 967-976. https://doi.org/10.1175/1520-0469(1985)042<0967:LOOTSA>2.0.CO;2.

[47] Kent, G. S., Ottway, M., Keenliside, W., Wright, R. W. H., Sandford, M. C. W., 1972: A study of feasibility of measuring atmospheric densities by using a laser-searchlight technique. AFOSR Report n. 43, Period: April 1964 – June 1971. Contract n. AF AFOSR-616-67. Available from: https://apps.dtic.mil/dtic/tr/fulltext/u2/741875.pdf. Accessed 13 October 2020.

[48] Clemesha, B., Kent, G. S., and Wright, W. H., 1966: Laser probing the lower atmosphere. Nature, v. 209, 184-186. https://doi.org/10.1038/209184a0.

[49] Grams, G., and Fiocco, G., 1967: Stratospheric aerosol layer during 1964 and 1965. J. Geophys. Res., v. 72, Issue 14, 3523-3542. https://doi.org/10.1029/JZ072i014p03523.

[50] Turco, R. P., Whitten, R. C., Toon, O. B., 1982: Stratospheric aerosols: Observation and theory, Rev. Geophys., v. 20, Issue 2, 233-279. https://doi.org/10.1029/RG020i002p00233.

[51] Simonich D.M., Clemesha B.R., 1997: A History of Aerosol Measurements at São José dos Campos, Brazil (23 S, 46 W) from 1972 to 1995. In: Ansmann A., Neuber R., Rairoux P., Wandinger U., editors. Advances in Atmospheric Remote Sensing with Lidar, Springer, Berlin, Heidelberg. https://doi.org/10.1007/978-3-642-60612-0_116.

[52] Simonich, D. M., Clemesha B. R., 1989: Decay of the El Chichón aerosol cloud at 23 S. J. Geophys. Res., v. 94, Issue D10, 12803-12806. https://doi.org/10.1029/JD094iD10p12803.

[53] Kremser, S., Thomason, L. W., von Hobe, M., Hermann, M., Deshler, T., Timmreck, C., et al., 2016: Stratospheric aerosol—Observations, processes, and impact on climate. Reviews of Geophysics, v. 54, Issue 2, 278-335. https://doi.org/10.1002/2015RG000511.

[54] Antuña, J. C., Estevan, R., Barja, B., 2012: Demonstrating the potential for

first-class research in underdeveloped countries: Research on stratospheric aerosols and cirrus clouds optical properties, and radiative effects in Cuba (1988-2010). Bull. Amer. Meteor. Soc., v. 93, Issue 7, 1017-1027. https://doi.org/10.1175/BAMS-D-11-00149.1.

[55] Antuña, J. C., Estevan, R., Barja, B., 2005: Características de los aerosoles en la troposfera alta y la estratosfera baja en el Gran caribe en ausencia de perturbación volcánica. Revista Cubana de Meteorología, v. 12, Issue 1, 65-72. (In Spanish). Available from: http://rcm.insmet.cu/index.php/rcm/article/view/310. Accessed 13 October 2020.

[56] Estevan, R., Antuña, J. C., 2010: Efecto radiativo de la erupción del Monte Pinatubo sobre Camagüey. Revista Cubana de Meteorología, v. 16, Issue 1, 90-98, (In Spanish), 2010. Available from: http://rcm.insmet.cu/index.php/rcm/article/view/136. Access 13 October 2020.

[57] Stenchikov, G. L., Kirchner, I., Robock, A., Graf, H-F., Antuña, J. C., Grainger, R. G., Lambert, A., Thomason, L., 1998: Radiative forcing from the 1991 Mount Pinatubo volcanic eruption. Journal of Geophys. Research, v. 103, Issue D12, 13837-13857. https://doi.org/10.1029/98JD00693.

[58] Antuña, J. C., Robock, A., Stenchikov, G. L., Thomason, L. W., Barnes, J. E., 2002: Lidar validation of SAGE II aerosol measurements after the 1991 Mount Pinatubo eruption. J. Geophys. Res., v. 107, Issue D14, ACL 3: 1-11. https://doi.org/10.1029/2001JD001441.

[59] Antuña J. C., Robock, A., Stenchikov, G. L., Zhou, J., David, C., Barnes, J., Thomason, L., 2003: Spatial and temporal variability of the stratospheric aerosol cloud produced by the 1991 Mount Pinatubo eruption. J. Geophys. Res. Atmos., v. 108, Issue D – 20. https://doi.org/10.1029/2003JD003722.

[60] Stevermer, A. J., Petropavlovskikh, I., Rosen, J. M., DeLuisi, J. J., 2000: Development of a global stratospheric aerosol climatology: Optical properties and applications for UV. J. Geophys. Res., 105, Issue D18, 22763-22776. https://doi.org/10.1029/2000JD900368.

[61] Lavorato, M., Cesarano, P., Quel, E., Flamant, P. H., 2002: A dual receiver-backscatter lidar operated in Buenos Aires (34.6 S / 58.5 W). Proc. 21th ILRC (International Radar Laser Conference), 75-78, Quebec, Canada.

[62] Estevan R., Antuña, J. C., Lavorato, M. B., 2008: Stratospheric aerosols measurements at CEILAP, Argentina: Two case studies. Opt. Pura Apl., v. 41, Issue 2, 101-107. Proceedings of the IV Workshop of Lidar Measurements in Latin America. ISSN-e 2171-8814. Available from: https://www.sedoptica.es/Menu_Volumenes/Pdfs/282.pdf. Accessed 13 October 2020.

[63] Salvador, J. O., Wolfram, E., Pallotta, J., Otero, L., D'Elia, R., Quel, E., 2008: Correction of a stratospheric ozone profile using stratospheric aerosols backscatter in Rio Gallegos, Argentina. A case study, Opt. Pura Apl., v. 41, Issue 2, 165-170. Available from: https://www.sedoptica.es/Menu_Volumenes/Pdfs/293.pdf. Access 13 October 2020.

[64] Organization of American States, 1990: Disaster, Planning, and Development: Managing Natural Hazards to Reduce Loss. Report. Washington, D. C. Available from https://www.oas.org/dsd/publications/Unit/oea54e/oea54e.pdf.

[65] Sigurdson, H., Houghton, B., McNutt, S., Rymer, H., and Stix, J., editors. The Encyclopedia of Volcanoes, 2nd Edition, 2015. Elsevier. Academic Press. ISBN: 9780123859389. https://doi.org/10.1016/C2015-0-00175-7.

[66] Lopes, F. J. S., Silva, J. J., Antuña, J. C., Ghassan, T., and Landulfo,

E., 2019: Synergetic Aerosol Layer Observation After the 2015 Calbuco Volcanic Eruption Event. MDPI. Remote Sensing, v. 11, Issue 12, 195. https://doi.org/10.3390/rs11020195.

[67] Bègue, N., Shikwambana, L., Bencherif, H., Pallotta, J., Sivakumar, V., Wolfram, E., Mbatha, N., Orte, F., Du Preez, D. J., Ranaivombola, M., Piketh, S., and Formenti, P., 2020: Statistical analysis of the long-range transport of the 2015 Calbuco volcanic plume from ground-based and space-borne observations, Ann. Geophys., v. 38, Issue 2, 395-420. https://doi.org/10.5194/angeo-38-395-2020.

[68] Watson, R. T., Geller, M. A., Stolarski, R. S., Hampson, R. F., 1986: Present State of Knowledge of the Upper Atmosphere: An Assessment Report. Processes That Control Ozone and Other Climatically Important Trace Gases. NASA Reference Publication 1162. Available from: https://ntrs.nasa.gov/citations/19860016439. Document ID: 19860016439. Accessed 2 November 2020.

[69] Farman, J., Gardiner, B., Shanklin, J., 1985: Large losses of total ozone in Antarctica reveal seasonal ClOx/NOx interaction. Nature 315, 207-210. https://doi.org/10.1038/315207a0.

[70] Pazmiño, A., Godin-Beekmann, S., Ginzburg, M., Bekki, S., Hauchecorne, A., Piacentini, R. D., Quel, E. J., 2005: Impact of Antarctic polar vortex occurrences on total ozone and UVB radiation at southern Argentinean and Antarctic stations during 1997-2003 period. Journal of Geophysical Research: Atmospheres, v. 110, Issue D3, D03103. https://doi.org/10.1029/2004JD005304.

[71] Mégie, G., Ancellet, G., Pelon J., 1985: Lidar measurements of ozone vertical profiles. Applied Optics, v. 24, Issue 21, 3454-3463. https://doi.org/10.1364/AO.24.003454.

[72] Wolfram E. A., Salvador, J., Otero, L., Pazmiño, A., Porteneuve, J., Godin-Beekmann, S., Nakane, H., Quel, E. J., 2005: SOLAR campaign: Stratospheric ozone lidar of Argentina. Proceedings of SPIE, v. 5887. Lidar Remote Sensing for Environmental Monitoring VI. 588713. https://doi.org/10.1117/12.620293.

[73] Wolfram, E. A., Salvador, J., D'Elia, R., Casiccia, C., Paes Leme, N., Pazmiño, A., Porteneuve, J., Godin-Beekman, S., Nakane H., Quel, E. J., 2008: New differential absorption lidar for stratospheric ozone monitoring in Patagonia, South Argentina. Journal of Optics A: Pure and Applied Optics, v. 10, Issue 10, 104021. https://doi.org/10.1088/1464-4258/10/10/104021.

[74] Wolfram, E. A., Salvador, J., Pallotta, J., D'Elia, R., Otero, L., Godin-Beekmann, S., Pazmiño, A., Nakane, H., Quel, E., 2006: SOLAR Campaign: first results of ozone profile measurements at Río Gallegos (51° 55' S, 69° 14' W), Argentina. The 23rd International Laser Radar Conference (ILRC23), Japan. Available from: https://laser-sensing.jp/ilrc23_CD1a2b3c/ILRC23/4O-4.pdf. Accessed 3 November 2020.

[75] Wolfram, E. A., Salvador, J., Orte, F., D'Elia, R., Godin-Beekmann, S., Kuttippurath, J., Pazmiño, A., Goutail, F., Casiccia, C., Zamorano, F., Paes Leme, N., Quel, E. J., 2012: The unusual persistence of an ozone hole over a southern mid-latitude station during the Antarctic spring 2009: a multi-instrument study. Annales Geophysicae, v. 30, Issue 10, 1435-1449. https://doi.org/10.5194/angeo-30-1435-2012.

[76] Ristori, P., Otero, L., Jin, Y., Barja, B., Shimizu, A., Barbero, A., Salvador, J., Bali, J. L., Herrera, M., Etala, P., Acquesta, A., Quel, E., Sugimoto, N., Mizuno, A., 2018: Saver.net lidar network in southern South America. The 28th International Laser Radar Conference (ILRC 28). EPJ Web Conf., v. 176, 09011. https://doi.org/10.1051/epjconf/201817609011.

Chapter 2

Lidar Observations in South America. Part II - Troposphere

Eduardo Landulfo, Alexandre Cacheffo, Alexandre Calzavara Yoshida, Antonio Arleques Gomes, Fábio Juliano da Silva Lopes, Gregori de Arruda Moreira, Jonatan João da Silva, Vania Andrioli, Alexandre Pimenta, Chi Wang, Jiyao Xu, Maria Paulete Pereira Martins, Paulo Batista, Henrique de Melo Jorge Barbosa, Diego Alves Gouveia, Boris Barja González, Felix Zamorano, Eduardo Quel, Clodomyra Pereira, Elian Wolfram, Facundo Ismael Casasola, Facundo Orte, Jacobo Omar Salvador, Juan Vicente Pallotta, Lidia Ana Otero, Maria Prieto, Pablo Roberto Ristori, Silvina Brusca, John Henry Reina Estupiñan, Estiven Sanchez Barrera, Juan Carlos Antuña-Marrero, Ricardo Forno, Marcos Andrade, Judith Johanna Hoelzemann, Anderson Guimarães Guedes, Cristina Tobler Sousa, Daniel Camilo Fortunato dos Santos Oliveira, Ediclê de Souza Fernandes Duarte, Marcos Paulo Araújo da Silva and Renata Sammara da Silva Santos

Abstract

In Part II of this chapter, we intend to show the significant advances and results concerning aerosols' tropospheric monitoring in South America. The tropospheric lidar monitoring is also supported by the Latin American Lidar Network (LALINET). It is concerned about aerosols originating from urban pollution, biomass burning, desert dust, sea spray, and other primary sources. Cloud studies and their impact on radiative transfer using tropospheric lidar measurements are also presented.

Keywords: lidar, LALINET, aerosols, atmospheric sciences, remote sensing, air quality, environment

1. Introduction

In Brazil, there are four lidar systems dedicated to the study of aerosols in the troposphere. Installed at Embrapa in Manaus (Western Amazon), there is a lidar system coordinated by the Atmospheric Physics Laboratory (LFA) of the Institute of Physics at the University of São Paulo [1]. In São Paulo, there are two other systems, the SPU Lidar Station, installed at the Institute of Energy and Nuclear Research (IPEN/CNEN) and coordinated by the Laboratory of Environmental Applications of Lasers (LEAL) hosted in the Center for Lasers and Applications (CELAP) of the referred institute, and the scanning lidar system located at CEPEMA (Centre for Training and Research in Environment) in Cubatão (State of São Paulo) [2–5]. **Figure 1** shows an example of the aerosol profile retrieved at SPU Station.

The DUSTER Lidar system, situated at the Department of Atmospheric and Climate Sciences (DCAC) at the Federal University of Rio Grande do Norte (UFRN), can measure marine aerosols' physical and optical properties. It can also measure aerosols (mineral dust) that cross the Atlantic Ocean and come from the desert Sahara, and aerosols originated from fires in the African continent [6, 7]. The lidar system's design and installation in Natal result from a technical and scientific collaboration among UFRN and IPEN. The DUSTER Lidar system is a biaxial monostatic lidar with a typical spatial resolution of 7.5 m. Brazil is a country with continental dimensions where different aerosols are generated, whether by natural or anthropogenic sources. The two systems mentioned above, SPU Lidar Station in São Paulo and DUSTER Lidar system in Natal, can measure different aerosols types.

Quality assurance and quality control programs developed by the European Aerosol Research Lidar Network (EARLINET) [8] are being implemented at the LALINET stations of São Paulo, Manaus, and Natal. This implementation intends to increase the capability to provide a reliable dataset in collaboration with three EARLINET stations (Bucharest, Granada, and Munich) in the framework of the project APEL (Assessment of atmospheric optical Properties during biomass burning Events and Long-range transport of desert dust) [9, 10]. The main objective is

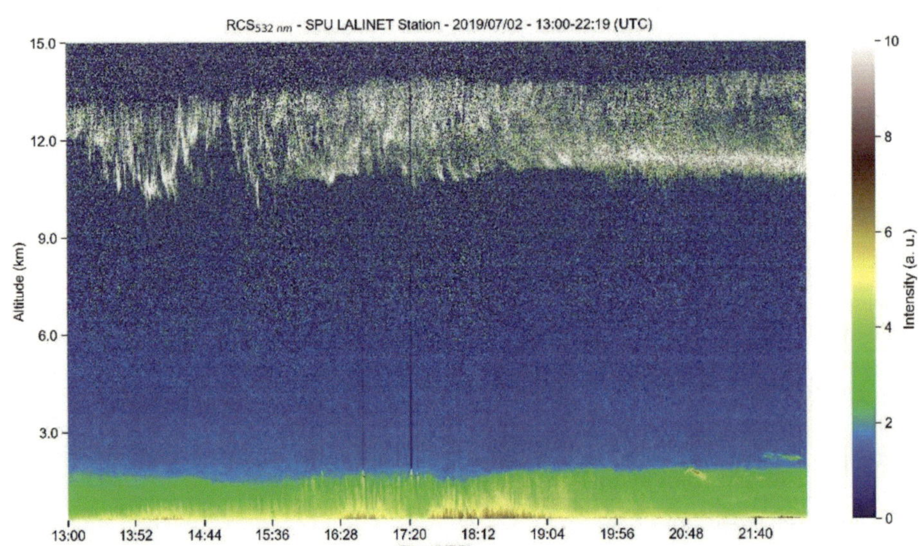

Figure 1.
Quick-look of the Lidar Range Corrected Signal (RCS) at 532 nm measured at SPU Lidar Station on 02 July 2019. The signal between 10 and 14 km indicates cirrus clouds.

to make the final data products from the two networks comparable and study the similarities and differences in aerosol loads, transport heights, types, and properties [11]. The evaluation will be done at the hardware and software levels. At the hardware level, the quality of the signals will be checked using the specific EARLINET procedures, and, at the software level, the LALINET data processing algorithms will be compared with the EARLINET Single Calculus Chain [12–14]. The last is a fully automatic evaluation process that can be used for virtually any lidar configuration and was validated for several EARLINET lidar stations, being a powerful tool that allows lidar stations to retrieve the aerosol backscatter and extinction profiles from the raw lidar data (**Figure 2**).

In Bolivia's case, the Laboratory for Atmospheric Physics of Universidad Mayor de San Andrés (LFA for its acronym in Spanish) is carrying out some studies related to urban aerosols and pollution monitoring in the metropolitan region of La Paz and El Alto. This region is one of the fastest-growing urban settlements in South America, with the particularity of being located in very complex terrain at a high altitude over the Andes. With a total population of around 1.8 million inhabitants is the second most populous urban area in Bolivia. La Paz city is located in a stepped valley, whose height starts at 3200 m a.s.l. (southern area), going up to 4000 m a.s.l. (in the north). The metropolitan area includes El Alto city (4100 m a.s.l), adjacent to the west's valley, and is extended over the Altiplano plateau. The valley has many basins that converge in the lower part of the city generating complex air fluxes.

An elastic lidar system was installed in the Science Campus of the Universidad Mayor de San Andrés (16.5333 S, 68.0667 W, 3420 m a.s.l.) in 2007 to study the boundary layer's behavior in this complex terrain. The lidar system was developed by improving an old system donated by the European Space Agency to the LFA and an essential collaboration of the Raman Lidar Laboratory of NASA's Goddard Space Flight Center. The instrument regularly worked for some years collecting data every Monday. These ancillary data were used for different short-term studies [15].

In 2011, this lidar acquired an additional relevance when a new Global Atmosphere Watch (GAW) station was set up near the metropolitan area at Mount

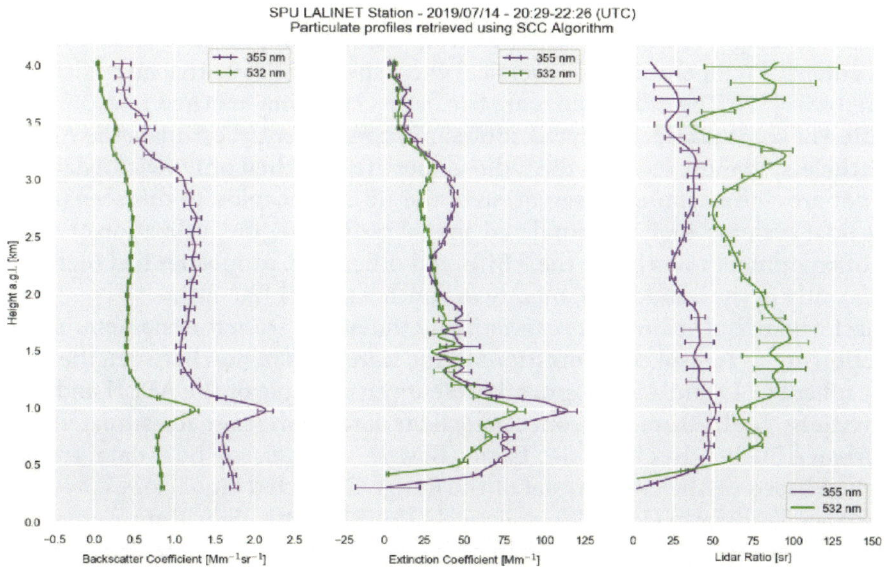

Figure 2.
Particle backscatter (Mm^{-1} sr^{-1}) and extinction (Mm^{-1}) coefficients and Lidar Ratio (sr), measured at SPU Lidar Station on 14 July 2019. Smoothed retrievals, obtained at 355 nm and 532 nm, using the Single Calculus Chain Algorithm.

Chacaltaya (16.3502 S, 68.1314 W, 5240 m a.s.l.). This station was set up to study aerosols' physical and chemical properties, measure atmospheric gas concentrations, study the aerosols injected into the free troposphere, besides the effect of aerosols deposition onto the Andean glaciers. In this sense, the main task of the lidar system was to help with the study of the air fluxes that go from the metropolitan area to the Chacaltaya GAW station and vice versa and the behavior of the local atmospheric boundary layer, especially in connection with atmospheric pollution in the urban area.

Besides, in 2018 and thanks to a collaboration of the Andalusian Institute for Earth System Research, Granada, Spain, a Lufft CHM 15 k ceilometer was installed in the northern part of La Paz city, closer to the Chacaltaya GAW station than the LFA. The goal was to characterize the boundary layer height's seasonal behavior through continuous measurements for at least one year. The Wavelet Covariance Transform (WCT) was used to estimate this behavior using both the ceilometer and the university campus's lidar. Although we gained knowledge about the local ABL's temporal behavior, it is clear that due to the complexity of topography in this region, extending this work's main conclusions is not straightforward. More measurements and modeling are needed for this purpose.

2. Urban aerosols and pollution monitoring

2.1 The atmospheric boundary layer

The Atmospheric Boundary Layer (ABL) is the lowest section of the troposphere and is directly affected by the surface, responding to surface forcing within a one-hour or less time scale. The ABL has turbulent properties and high variability in its daily cycle, and it is a fundamental parameter to several studies, e.g., air quality, numerical weather forecasting, climate modeling, and wind energy applications [16]. These characteristics, associated with the variations in the ABL stability, enable us to subdivide it into three main layers: The Convective Boundary Layer (CBL), the Stable Boundary Layer (SBL), and the Residual Layer (RL).

The ABL height (ABLH) is obtained from the vertical profile of some tracers like a potential temperature [17], vertical wind speed [18], relative humidity [19], and aerosols [11]. The radiosondes are the more traditional method to estimate the variation of some tracers indicated above and, consequently, estimate the ABLH. Nevertheless, in most regions, the radiosondes are launched only twice a day, which does not provide a detailed observation of the ABLH behavior. In this scenario, due to the lidar systems' high temporal and spatial resolution, the utilization of this kind of equipment to estimate the ABLH and other ABL properties had increased significantly in the last decade, mainly in South America [20–30].

Elastic lidar and ceilometers can estimate the ABLH from the characteristic reduction in the aerosol concentrations in the transition region between the Free Troposphere (FT) and ABL. **Figure 3** presents an example of the ABLH and its subdivisions, both estimated from elastic lidar data. Moreira et al. estimated the Urban ABLH to the city of São Paulo (Brazil), using elastic lidar data, from a method based on the curtain-plot of the Range Corrected Signal and Wavelet Covariance Transform (WCT), respectively [31, 32]. Both algorithms were validated by radiosonde data, resulting in high correlations during convective and clear sky conditions. Also, based on WCT, Niesperuza et al. estimated the ABLH to Medellín (Colombia), demonstrating the influence of the selections of the parameters in the Haar Wavelet performance [33]. Salvador et al. performed a comparison among the ABLH estimated from elastic lidar, SODAR, and Weather Research and

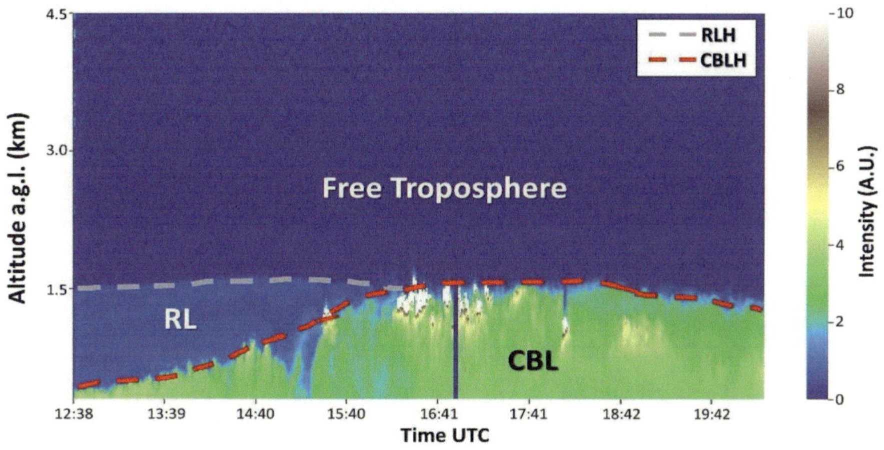

Figure 3.
ABL and its subdivisions. WCT and Gradient Methods, applied in elastic lidar data, estimated the CBLH and RLH. The aerosol profiles were measured at SPU Lidar Station on 03 August 2020.

Forecasting Model (WRF), finding a high positive correlation during the convective period [26]. Besides, elastic lidar data can also observe aerosol plumes' movement as the mixing level in the CBL region, from the skewness and kurtosis profiles. This method was applied in the city of São Paulo [34]. The methodology can be applied in studies about air quality providing a better observation about pollutant concentrations.

In comparison with elastic lidar, Doppler lidar provides more possibilities to identify the ABLH due to the capacity to obtain the wind speed profile, which can be applied as a tracer from its variance. Using Doppler Lidar data, Moreira et al. estimated the ABLH from the variance in the wind speed profile in São Paulo [24]. This method was compared with radiosonde data demonstrating a high correlation in CBL and SBL situations. The wind speed profile was used to detect low-level jets (LLJ). Then from the maximum of LLJ, the SBL height was estimated [25]. Marques et al. used the maximum variance in the Noise Ratio to estimate the ABLH. Such a result was compared with radiosonde data, reaching high correlations in stable and convective situations [28].

2.2 Retrievals from the LiDAR-CIBioFi station at Cali-Colombia

In Colombia (Cali), to detect the ABL altitude, lidar signals obtained from the LiDAR-CIBioFi station at Universidad del Valle are employed. The study site is georeferenced in **Figure 4**. The methodology employs an interplay between the Gradient [32, 33] and WCT [34] Methods, as described in detail in Ref. [27].

Statistical validation of the implemented instrumentation is performed to support the data quality by contrasting atmospheric profiles retrieved by radiosondes launched at the local international airport, a few kilometers away from our station. The maximum vertical gradient level of potential temperature is used to detect the ABL top (ABLT) by employing radiosonde profiles. A linear relationship between the daily ABLT evolution retrieved by the lidar station and the radiosonde profiles goes as follows: ABLLiDAR = 0.967 × ABLRadio - 0.022. It is statistically significant at the 95% confidence level and R^2 (consider the separation between the radiosonde launching site and the lidar station, see **Figure 4b**).

Once the data are validated, the ABLT levels are compared against the available local Particulate Matter (PM) concentration information. The correlations between

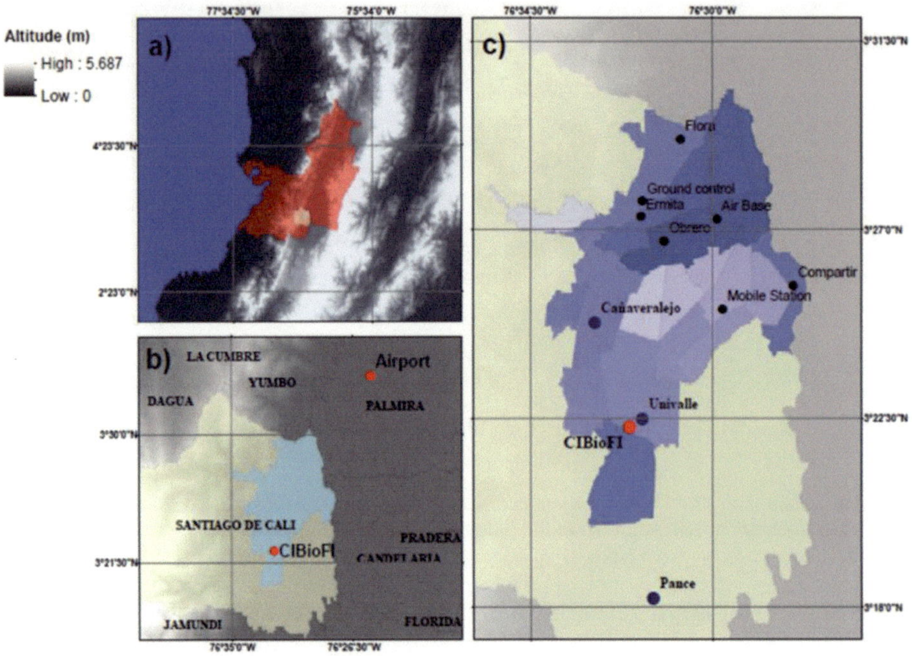

Figure 4.
(a) The location of Valle del Cauca county (the gray area) and the city of Cali (white area) in Colombia (altitude values according to the color scale). (b) The light color shape represents Cali's rural area. The blue region is linked to the city's urban area. The red dots mark the LiDAR-CIBioFi system's location at Universidad del Valle (UniValle) and the radiosonde station (at Cali's international airport). (c) Political and administrative division of Cali. The blue dots place the Air Quality Monitoring System (SVCA) ground stations of the local Administrative Department of Environmental Management (DAGMA).

daily ABLT evolution and PM concentration data from three representative city ground-stations (Cañaveralejo, Pance, and UniValle), shown in **Figure 4c**, are analyzed. A strong negative relationship for the Cañaveralejo station gives $R^2 = 0.79$, while the Pance station exhibits an unencouraging positive slope with a correlation coefficient $R^2 = 0.13$, meaning a PM concentration increase for higher ABLT values, with PM values above the World Health Organization limits. The UniValle station, located about 100 m away from the LiDAR-CIBioFi station, reveals a low negative correlation ($R^2 = 0.20$) for the ABLT evolution for all months, especially at the beginning of the wet season.

An innovative method for retrieving the ABL top from LiDAR signals was developed at the LiDAR-CIBioFi station. It consists of training a convolutional neural network (NN) in a supervised manner, driving it to learn how to retrieve this dynamical parameter on real, non-ideal conditions and, in a fully automated and unsupervised process [31]. The Wavelet Covariance Transform (WCT) is used as a labeling method for constructing the training data set and as a baseline method for comparison with the trained NN.

The dataset used for the model's training and tuning is composed of 15,000 signals extracted from daytime measurements taken during December 2018 and February 2019. The signals were labeled using WCT, with a custom search threshold for each one; this was done to ensure the labels' quality and, consequently, the neural network predictions [31].

It is expected that the corrected training of the model replicates the predictions of a signal-by-signal fine-tuned WCT but in a completely automated and non-supervised process. The convolutional neural network proposed for the ABLT detection is compared to WCT in a supervised variant (custom search threshold for all time evolution) and unsupervised WCT (full signal as input during all time evolution) [31]. WCT was chosen as the labeling and comparison method for its ease of implementation and well-known robustness and performance, as for the past two decades, it has been used to measure the ABLT in numerous case studies [31].

On 14 August (**Figure 5a**), clouds around 4 km height, with some formations around 2 km, were detected, with the latter being very close to the boundary layer's height, thus posing a challenge for accurate ABLT detection. Besides, some cases of changes in density were detected after 14 h [31].

The first single lidar measured profile (**Figure 5b**), taken at about 12:30 h, exhibits a well-mixed layer, making it easy to discriminate between ABL and free troposphere. This condition allows a straightforward evaluation of the predictions: the NN gives very similar results to the supervised WCT (about 1.8 km), while the unsupervised WCT located the ABLT in a cloud formation above 4 km height [31]. In contrast, the second profile (**Figure 5c**), taken at 15:40 h, gives very different results for the supervised (sup.) WCT, the unsupervised WCT, and the NN. The supervised WCT located the ABLT at 500 m, below the expected result.

The unsupervised WCT placed the result at around 4 km, in a cloud formation pattern, while the NN located the ABLT about 2.6 km, following its actual behavior [31]. The temporal evolution of the LiDAR measurement profiles of **Figure 5** clarifies that the unsupervised WCT detection locates the boundary layer position at cloud formation height, erroneously placing the ABLT in most cases [31]. The supervised WCT shows ABLT detection problems, severely underestimating ABLT for cases of proximity to clouds (see, e.g., around 12 h), and in cases of residual layers (e.g., after 14 h). Despite the drawbacks of supervised WCT and unsupervised WCT, **Figure 5** clearly shows that our convolutional NN estimation of ABLT is more resilient to nearby clouds than WCT [31].

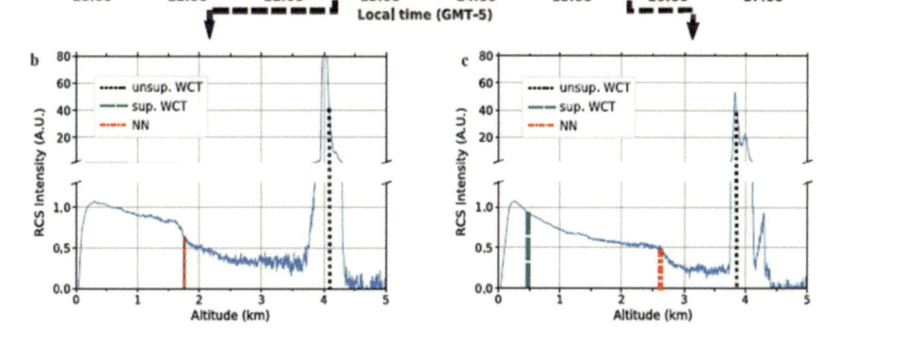

Figure 5.
Lidar retrievals for 14 August 2019 [30]. (a) Temporal evolution of the ABL, and (b), (c) Two selected single profiles. They point out different scenarios treated with our method: the measurements exhibit b Profile 1 (12:30 h), a well-mixed layer where NN and sup. WCT values are very similar, and c, Profile 2 (15:40 h), shows conditions where the NN estimation differs from supervised and unsupervised WCT values; the latter profile exhibits an extended ABL a height of about 2.6 km for the NN prediction. The intensity of the signals is given in arbitrary units (a. u.) [31].

The SVCA is the local Air Quality Monitoring System, a governmental policy for continuously monitoring air quality and assessing the pollutants trend. This air quality network provides information in the medium and long term to support developing strategies to attend to the air pollution effects from a holistic view. This network is composed of nine ground stations distributed, as shown in **Figure 4**. There are six stations to measure PM_{10} concentration and four stations for $PM_{2.5}$ concentration measurements. These stations count with automatic analyzers technology for monitoring aerosols and gases such as ozone, carbon dioxide, and others. Employing the raw data retrieved by these stations, the local Administrative Department of Environmental Management (DAGMA) makes monthly and annual reports, which can be found [31] at http://www.cali.gov.co/dagma.

Hourly raw data from three ground base stations that belong to the SVCA network were used to analyze the daily behavior of PM_{10} and $PM_{2.5}$. Due to its proximity to the LiDAR-CIBioFi station, Cañaveralejo, Pance, and UniValle stations have been picked. Specifically, the UniValle station is about 100 m away from the LiDAR-CIBioFi, and for a direct intercomparison with ABL altitude is a primary source of data. The data retrieved from these ground stations are significant since they are the only available aerosols data source that features Cali to support environmental public policy reinforcement; these are meant to control day-by-day vehicular fleet restrictions and industrial emissions.

One way to assess the impact of PM concentration on public health and radiative forcing is to study its response according to the ABL vertical and horizontal dynamics. We account for PM_{10} and $PM_{2.5}$ concentrations retrieved by the automatic analyzers and the vertical atmospheric profiles from LiDAR-CIBioFi. The correlations for the PM as a function of the ABL altitude retrieved by the LiDAR system, the vertical response of PM for each station, according to the ABL daily evolution during July, August, September, and October 2018, are analyzed. The daily data are used as a first step to identify PM daily behavior within the ABL.

The hypothesis to understand the relationship between PM and ABL is that the aerosol's mass volume has an inverse behavior to the ABLT evolution. **Figure 6** shows how daily data from Cañaveralejo station (red dots) already comply with this hypothesis: PM_{10} concentration decreases with the ABL height; this behavior is a continuous trend for each of the four covered months, regardless of the temperature and solar radiation transition along the day. Especially in the mornings, the Cañaveralejo station reports the maximum concentration amongst the three stations, with values not above the local 24-hour limit.

Regarding Pance (green dots) and UniValle (blue crosses) stations, the regressions exhibit odd behavior, particularly during July and August 2018: **Figure 6a** and **b** show that Pance's PM_{10} concentration dangerously grows with increasing ABL height, a situation of concern since this means that the population remains exposed to high concentrations within a large air volume mass in the ABL. Since we do not have additional information about the emission source, we postulate that Pance high concentrations could be attributed to tropical forests' haze. Nevertheless, it is necessary to warn on this behavior and spot possible emission sources, and identify such aerosols' optical properties.

The Pance and Cañaveralejo stations were chosen due to their proximity to the LiDAR-CIBioFi station, but their locations correspond to a different landscape, and the ABLT dynamics could be very different from that of UniValle's. Since UniValle station is located about 100 m away from the LiDAR-CIBioFi station (and within the same university campus), it is a direct source for PM data comparison/analysis against ABLT dynamics; it is, however, necessary to keep in mind that UniValle station only makes measurements of fine ($PM_{2.5}$) particles concentration. Even though UniValle station shows $PM_{2.5}$ values well below those of PM_{10} Pance and

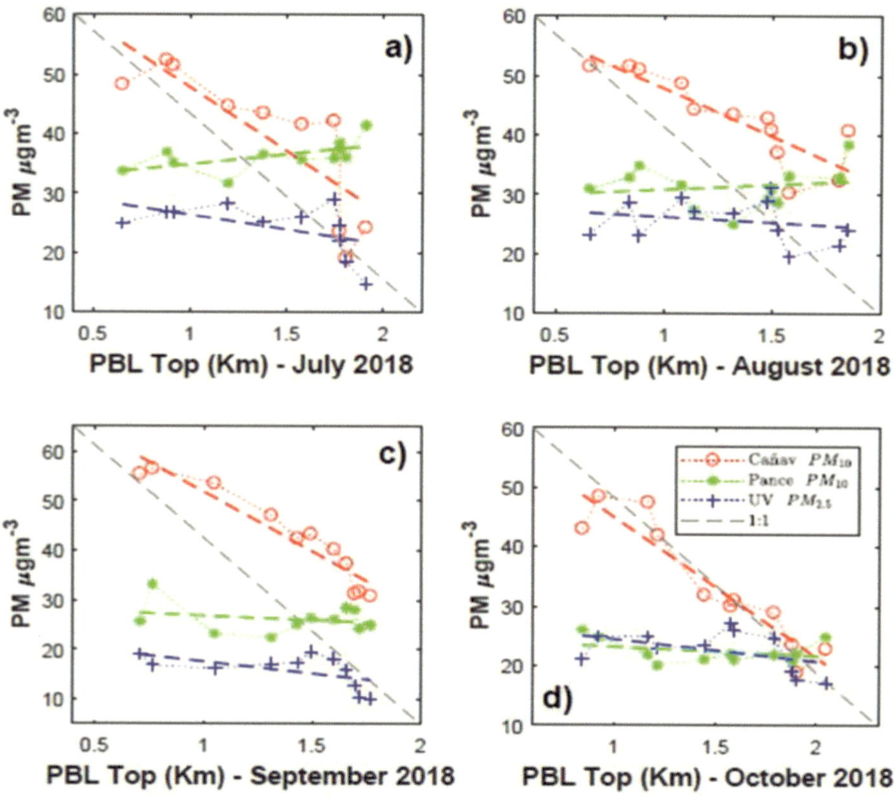

Figure 6.
(a) Linear regression between PM (µg m⁻³) concentration and ABLT (or PBL top-planetary boundary layer top) altitude retrieved by LiDAR-CIBioFi in July 2018. (b), (c), and (d) denote the same as in (a) except for the month of data collected (August, September, and October 2018, respectively).

Cañaveralejo stations during July, August, and September, during October 2018, Pance and UniValle registered very similar values for $PM_{2.5}$ and PM_{10} against the ABL height; this means a high worrying indicator of fine ($PM_{2.5}$) particles concentration at UniValle. The linear regression between UniValle and Pance stations and the ABL height shows similar behavior. Thus, $PM_{2.5}$ concentration values are kept almost constant during the whole day and ABL height evolution. These fine particles' size allows them to stay longer in the atmosphere, explaining the regression line trend.

All data for each station in **Figure 6** and corresponding regressions were organized chronologically. Hence, regarding air quality, unquestionably, the most critical report is that for UniValle station due to the size of measured particles and the worrying linear regression balance with the ABLT daily evolution. The reported concentration levels are slightly above the local annual threshold and the World Health Organization (WHO) 24-hour PM limits.

To summarize, the ABLT results obtained at the LiDAR-CIBioFi station were compared against three PM SVCA ground-stations (Pance, Cañaveralejo, and UniValle) to analyze the behavior of the correlation with PM concentration near the LiDAR station. The Cañaveralejo station that carries out PM_{10} measurements shows an inversely proportional relationship with the ABLT, indicating that the population is not overexposed to PM_{10} concentration as higher ABLT values are reached. On the other hand, PM_{10} and $PM_{2.5}$ concentrations retrieved by Pance and UniValle stations show a different relationship with ABLT. Unexpectedly, the linear correlation slopes for each one of these stations were quite close to zero. The slope's mean value for

Pance station was 1.67, reporting positive values for July and August, which means that PM_{10} concentrations increased with increasing ABLT, a word of warning due to a possible negative impact on public health. On the other hand, the slope's mean value for UniValle ($PM_{2.5}$) station was −3.85, with negative monthly mean values.

3. Detection of biomass burning events

Wildfires generate large amounts of suspended particles in the atmosphere and increase the levels of carbon monoxide. The presence of these particles reduces both visibility and solar radiation reaching the Earth's surface. Besides, they act as cloud condensation nuclei, modifying the climate and the air composition and being harmful to human health [35]. On 8 November 2019, a dense feather of smoke was detached from Australia's coasts due to the intense fires that affected the region. These smoke layers were dragged by the winds to South America, entering Argentine territory on 14 November and remaining in suspension until the next day. In the particular case of the fires in Australia, a large amount of soot not only affected the entire surrounding region, devastating forests, and wildlife: the effects were seen around the planet, with measurements of the transport of aerosols at thousands of kilometers from the emission sources. An immediate effect of the accumulated soot from such a biomass burning was the alteration of river courses and the drinking water production in Eastern Australia [36].

The Australian fires started in September 2019 and intensified in November, given the drought conditions that affected the region. It was the second warmest summer registered, having a rainfall regime below the Australian summer average [36]. These major fires produced dense smoke plumes, detected by the Suomi NPP (National Polar-orbiting Partnership) VIIRS (Visible Infrared Imaging Radiometer Suite) satellites. The images are presented in **Figures 7** and **8** for 8 November, and in **Figures 9** and **10**, for 9 November. The figures show Australia's east coast, the most affected area. **Figures 7** and **9** show the Earth's surface's natural-looking satellite images, called True Color RGB images (I1-M4-M3). Meteorological clouds can be distinguished in white and smoke layers in translucent gray tones. **Figures 8** and **10** show another combination of spectral bands (M11-I2-I1), which allows observing, in shades of blue, the smoke plumes and, in reddish shades, the scars left on the surface of the Earth by fires (burned surface) [37]. These smoke plumes crossed the Pacific Ocean, reaching the American continent and Argentine territory in mid-November 2019.

Measurements from sensors onboard satellites and ground-based platforms were used to analyze the November biomass burning aerosols intrusion event from Australia. Within the satellite measurements, the data from the OMPS sensor (Ozone Mapping Profiler Suite) [38, 39] onboard the Suomi NPP satellite were analyzed for the study of the space–time variability of the Aerosol Index (AI Aerosol Index). AOD (Aerosol Optical Depth) measurements at 550 nm were retrieved from the MODIS (Moderate Resolution Imaging Spectroradiometer) instrument onboard the TERRA satellite [40, 41]. Additionally, the total columnar CO (carbon monoxide) content and the AI were sensed by the TROPOMI (TROPOspheric Monitoring Instrument) instrument onboard the Sentinel-5P satellite [42, 43]. AI is a qualitative index that indicates aerosol's presence at the higher layers of the atmosphere, absorbing or reflecting UV radiation. The main types of aerosols detected with this index are desert dust, biomass burning, and volcanic ash plumes. An advantage of AI is that it can be calculated for clear and (partially) cloudy ground pixels.

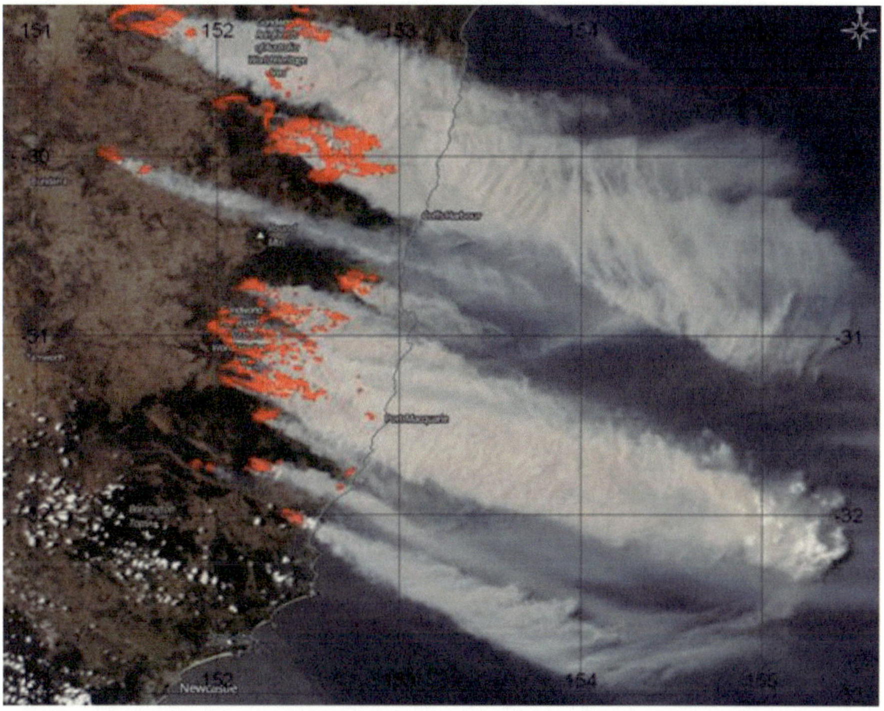

Figure 7.
Satellite image of Australia's east coast from 8 November 2019. VIIRS - Suomi NPP sensor (combination: I1-M4-M3).

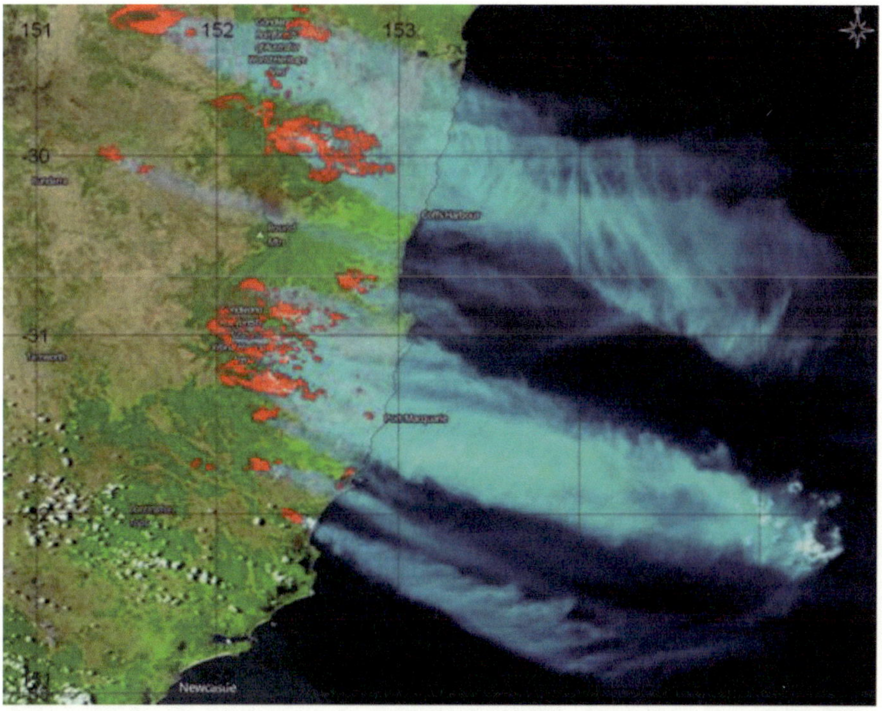

Figure 8.
Satellite image of Australia's east coast from 8 November 2019. VIIRS - Suomi NPP sensor (combination: M11-I2-I1).

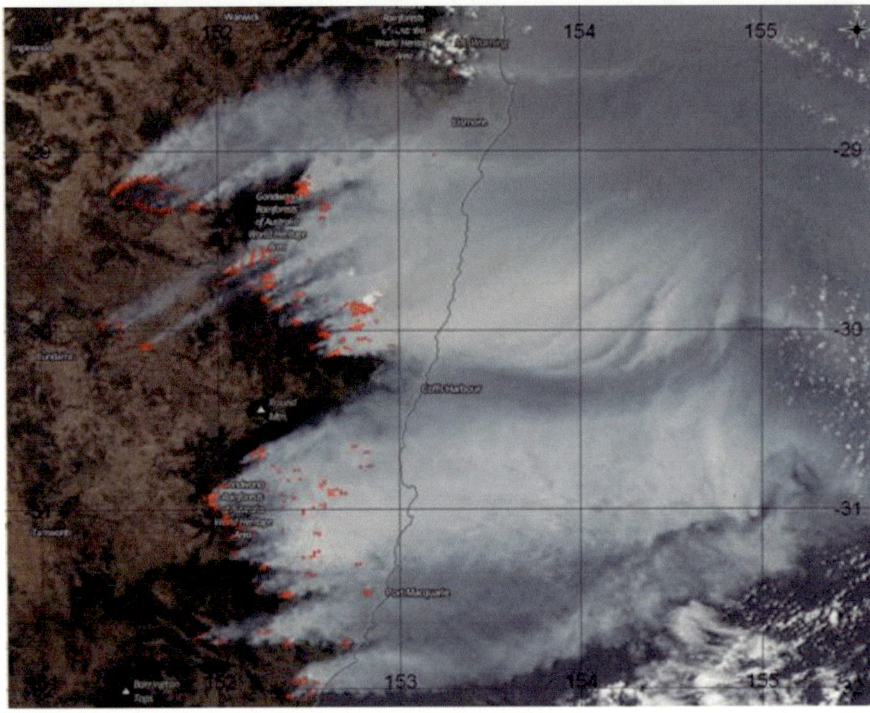

Figure 9.
Satellite image of Australia's east coast from 9 November 2019. VIIRS - Suomi NPP sensor (combination: I1-M4-M3).

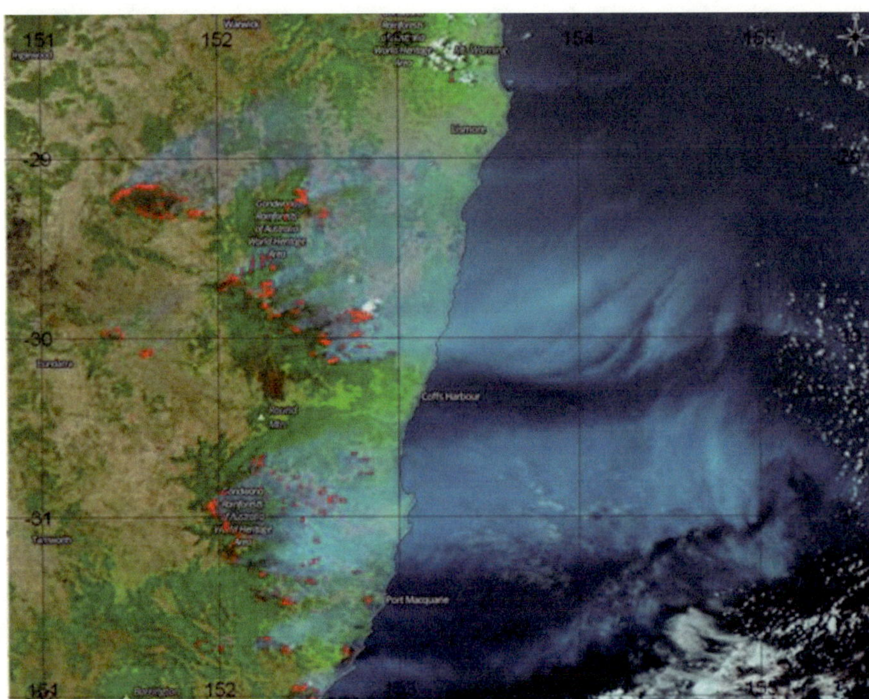

Figure 10.
Satellite image of Australia's east coast from 9 November 2019. VIIRS - Suomi NPP sensor (combination: M11-I2-I1).

The ground-based aerosol optical depth (AOD) data is obtained from the AERONET/NASA sun photometer network measurements at Buenos Aires, CEILPA-BA (34.555 W, 58.506 S, 26 m), Córdoba, and Pilar (31.667 W, 63.883 S, 333 m) stations, at Level 1.5 (data where the clouds have been extracted automatically) [44, 45].

The AOD is the aerosol vertical column integrated extinction at a given wavelength. This dimensionless quantity indicates how much aerosols attenuate the solar radiation as it passes through the atmosphere. Another value, the Ångström coefficient (or exponent), shows the AOD spectral dependence, and it is related to the root mean square distribution of the aerosol radii. It is calculated as the slope of the linear fit of the spectral AOD in a particular wavelength interval in a log–log scale graph. By relating the AOD at 440 nm and the Ångström coefficient, it is possible to classify the aerosol type using the classification table of Reference [46].

Measurements from the lidar instrument installed at CITEDEF were analyzed to determine the height of the aerosol layers. The normalized aerosol backscattering coefficient was calculated at 532 nm [47–49]. This system allows measuring the atmosphere's profiles from a few meters to several kilometers, exceeding the tropopause height up to the lower stratosphere. **Table 1** summarizes the variables analyzed, the sensors, and the platform employed.

Figure 11 shows the AI's space–time evolution measured by the OMPS sensor from 8 to 13 November 2019. The images show how a high AI value (greater than 5) smoke plume emerges from Australia's coasts on day 8 November 2019. This plume advances over the Pacific Ocean and reaches the coast of South America on 13 November 2019.

On 14 November 2019, the Australian smoke arrived over Argentine territory for the first time, through Neuquén province and covering the country's entire central region. **Figure 12** shows the AI coverage map (OMPS), indicating aerosols' presence throughout the area.

Figure 13 shows the TROPOMI sensor AI measurement for 14 November. The AI retrieved from OMPS and TROPOMI show similar values, around 1, in almost the entire territory and, in particular, values between 2 and 3, in the province of Entre Ríos. Both overlapping measurements are plotted in **Figure 14** to compare the AI measurements coverages with the two sensors. The TROPOMI measurement is taken as the basis, and the polygon (red outline) from the OMPS AI coverage is superimposed. It can be seen that the presence of aerosols in the upper layers of the atmosphere is the same for both sensors. It is known that biomass burning is one of the primary sources of CO release to the atmosphere. **Figure 15** shows the regional

Platform	Sensor	Variable
Satellite Suomi NPP	OMPS	AI
Satellite Sentinel 5p	TROPOMI	CO AI
Satellite TERRA	MODIS	AOD (550 nm) Ångström coefficient
Ground-based	LIDAR	Aerosol backscatter
Ground-based	Sun Photometer	AOD (440 nm) Ångström coefficient

Table 1.
Instruments and variables used to analyze the November biomass burning aerosols intrusion event.

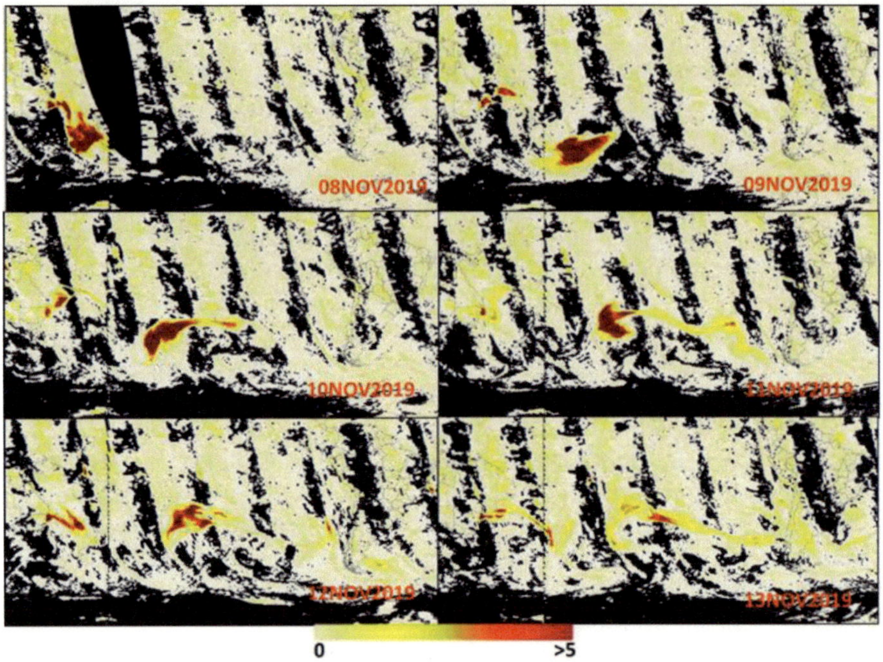

Figure 11.
Spatial–temporal evolution of the smoke plume through the OMPS sensor analysis of the Aerosol Index (AI), from 8 to 13 November 2019.

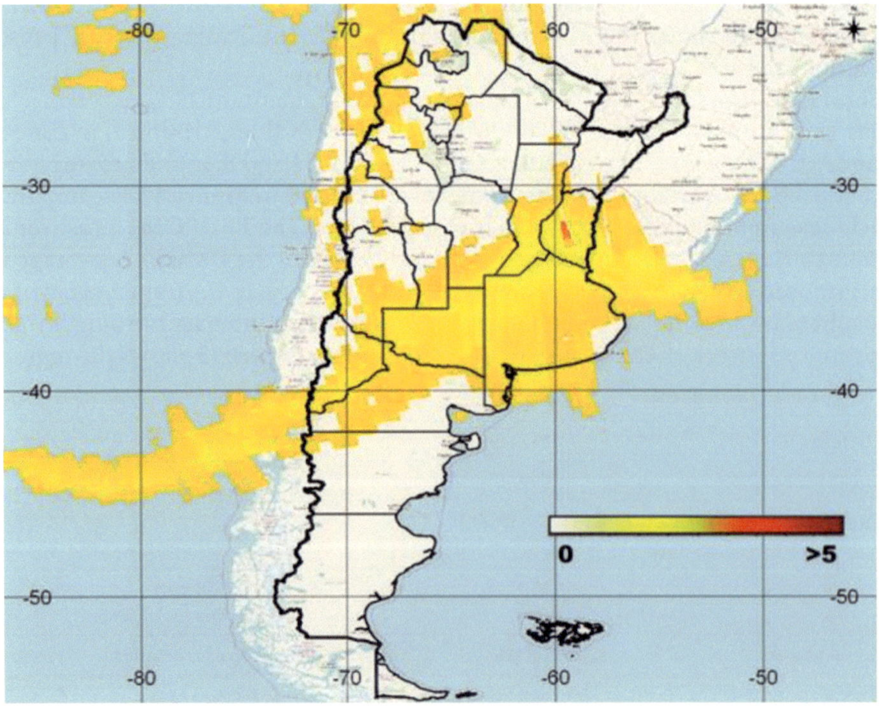

Figure 12.
Aerosol Index (AI), calculated by the OMPS sensor measurements for 14 November 2019.

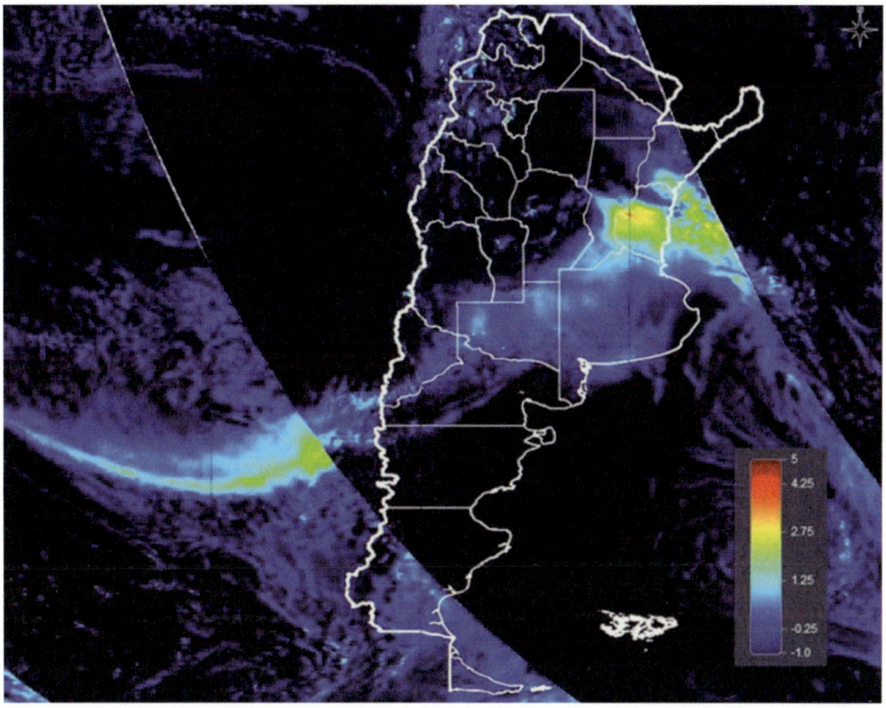

Figure 13.
Aerosol Index (AI), measured with the TROPOMI sensor for 14 November 2019.

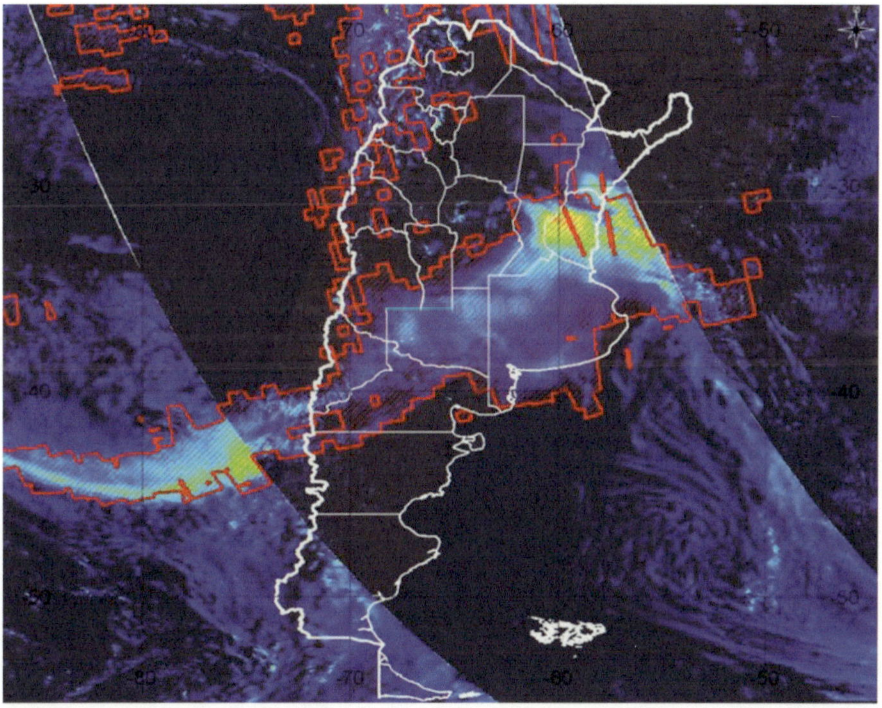

Figure 14.
Comparison of the Aerosol Index (AI) coverages, calculated with the TROPOMI sensor (in color palette) and the OMPS sensor (polygon in red), for 14 November 2019.

total columnar abundance of CO measured by the TROPOMI instrument [50]. The values observed throughout the region are not high.

A slight increase of CO measurements was observed in the same area where AI values were high, as shown in **Figure 16**, where the CO measurement overlaps the AI coverage (OMPS - red polygon). CO measurements reached a maximum of 0.04 mol/m^2 over the province of Entre Ríos.

A slight increase of CO measurements was observed in the same area where AI values were high, as shown in **Figure 16**, where the CO measurement overlaps the AI coverage (OMPS - red polygon). CO measurements reached a maximum of 0.04 mol/m^2 over the province of Entre Ríos.

Another interesting measurement to analyze the aerosols' presence is the AOD, which indicates at which level the aerosols prevent the sunlight from passing through the atmosphere. Aerosols scatter and absorb sunlight, resulting in reduced visibility. An AOD of less than 0.1 is characteristic of a clean atmosphere, with a very low number of suspended particles and maximum visibility. The AOD increases due to the increase of suspended particles, and this causes visibility loss.

Figure 17 shows the AOD measurement at 550 nm from the MODIS - TERRA sensor for 14 November 2019. Maximum values of 0.6 are observed in the southern part of the province of Entre Ríos. Values about 0.3 in the vicinity of the City of Buenos Aires and 0.2 in the vicinity of the City of Córdoba are also observed.

This satellite measurement can be contrasted with the AERONET/NASA sun photometer measurements available at Buenos Aires and Córdoba. **Figure 18** shows the AOD temporal evolution at 440 nm for 14 and 15 November 2019 for the Buenos Aires station and 15 November 2019 for the Córdoba station. At both stations, the values are higher than 0.1 along the two days.

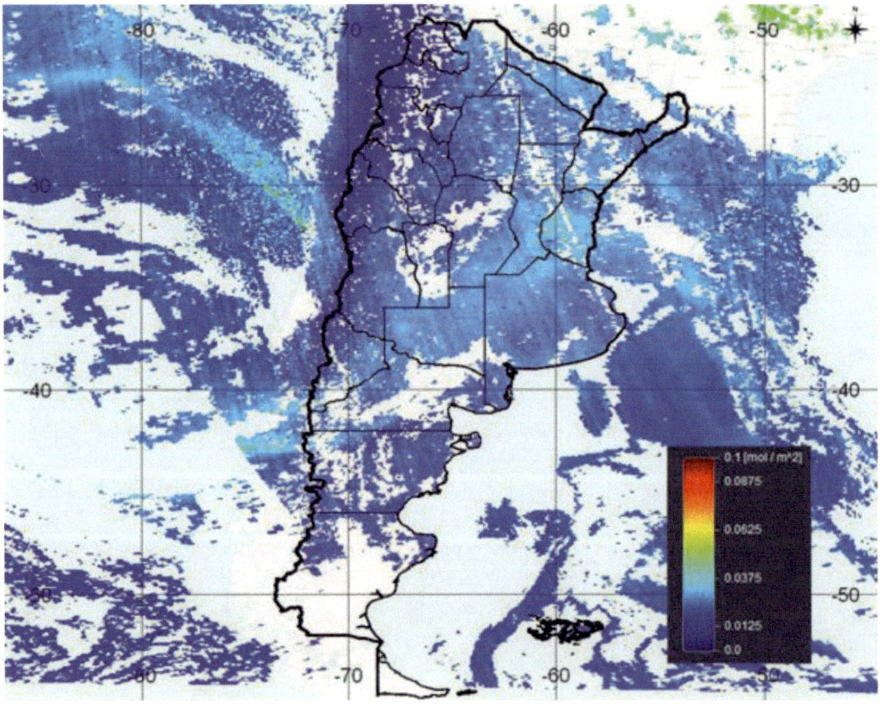

Figure 15.
CO in the total column from TROPOMI-Sentinel-5P for 14 November 2019.

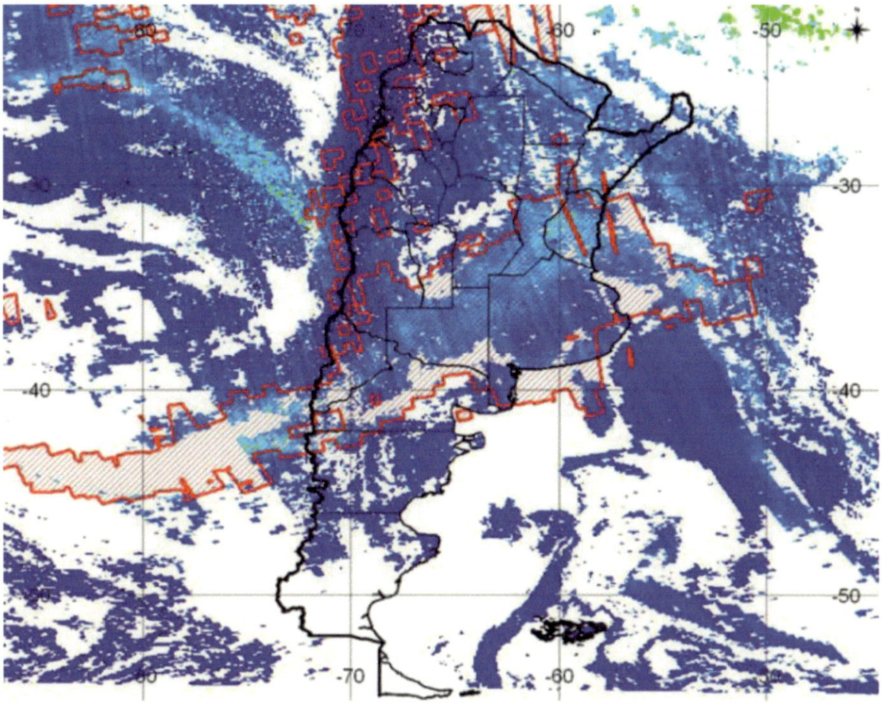

Figure 16.
CO in the total column from TROPOMI-Sentinel-5P and the coverage of the AI (OMPS - red polygon).

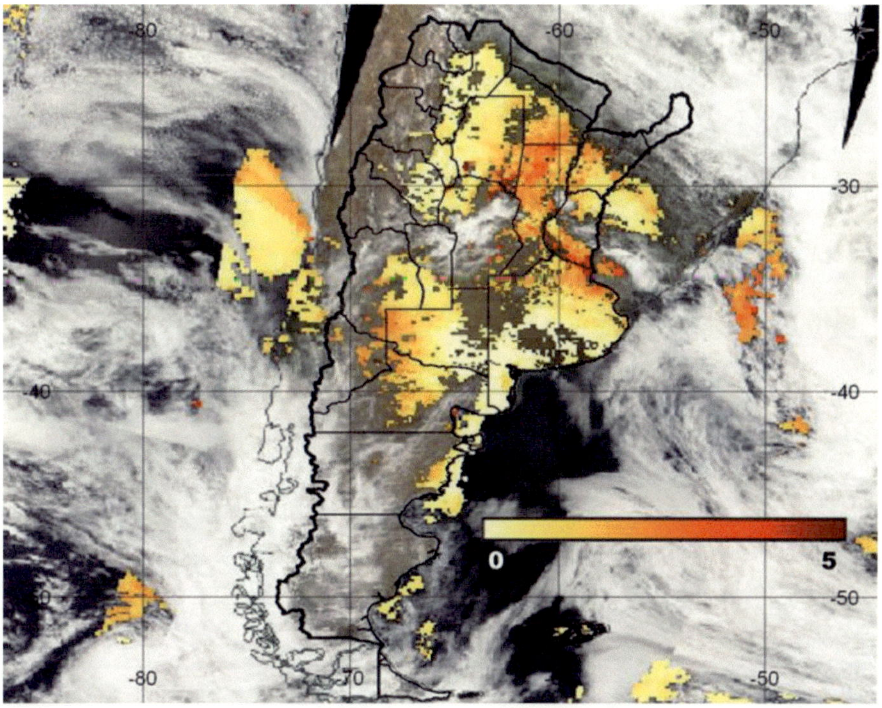

Figure 17.
AOD at 550 nm from MODIS-TERRA for 14 November 2019 (logarithmic scale).

The Ångström coefficient calculation can be used to provide additional information related to the size of the aerosols. The higher this coefficient is, the smaller the particle size. Values less than 1 suggest the domain of coarse particles (e.g., dust), and values greater than 1 suggest the domain of fine particles (e.g., smoke). **Figure 19** shows the Ångström coefficient from MODIS-TERRA product, where maximum values of 1.8 are observed in the southern part of Entre Ríos (in light blue), and 1.1 in the surroundings of Buenos Aires and south of Santa Fe (in green).

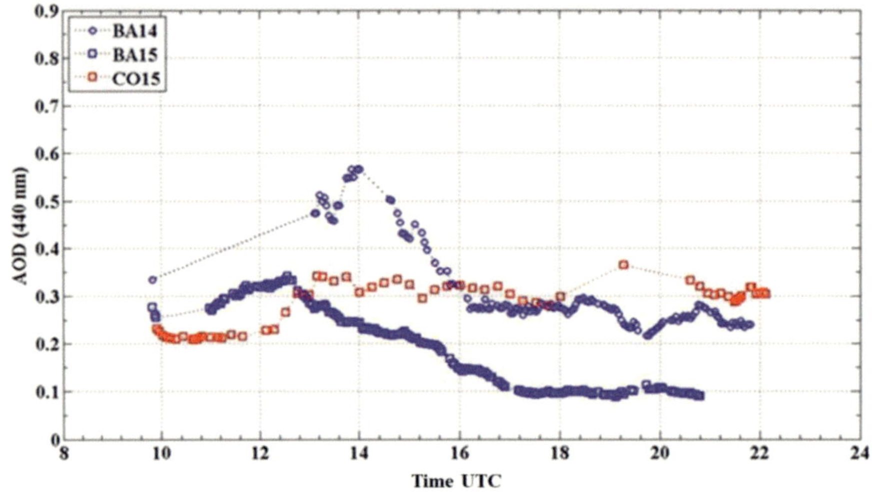

Figure 18.
Temporal evolution of the AOD at 440 nm during 14 and 15 November 2019 (BA: CEILAP-BA station (blue); CO: Pilar station (red)).

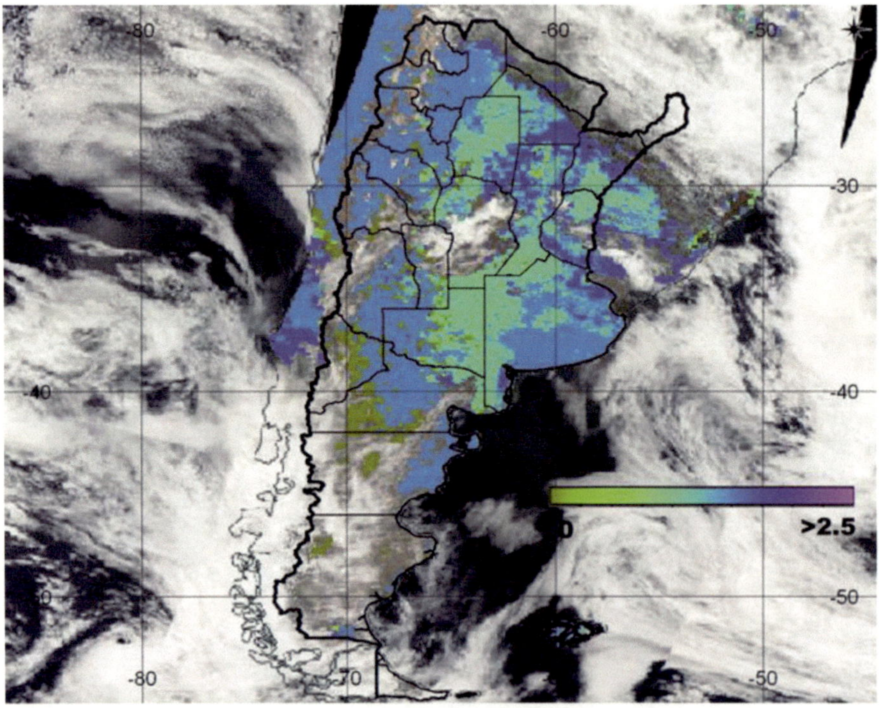

Figure 19.
Ångström coefficient (Blue) of MODIS-TERRA for 14 November 2019.

The Ångström coefficient was calculated from the AERONET/NASA sun photometers' AOD data at 870, 670, 500, and 440 nm. **Figure 20** shows the AOD at 440 nm versus the calculated Ångström coefficient. This style of graphs allows classifying the type of aerosols suspended in the atmosphere. From such an analysis, it can be extracted that most of the particles are originated from biomass burning, therefore being of the types Biomass Burning and Contaminated Continental, according to the classification table in Ref. [46].

Measurements with the lidar system made it possible to determine the aerosol layers' heights over the City of Buenos Aires. **Figures 21** and **22** show the normalized aerosol backscattering spatial–temporal evolution at 532 nm. On the horizontal axis is UTC's time (Local Time is UTC-3 h); on the vertical axis is the height in kilometers (the tropopause is about 13 km). The color palette shows the intensity of the signal. The blue color represents a clean, molecular atmosphere (without suspended particles), and the red color indicates particulate material. For both days, numerous well-defined layers of aerosols were observed above the atmospheric boundary layer at various heights and with different intensities, up to 13 km.

This work analyzed one event of the arrival of smoke plumes from Australia's intense fires to the Argentinean territory in November 2019. The study determined that, during that period, the AOD values were about 0.25, on average, and the Ångström coefficients were about 1.2. The aerosol layers were found above the atmospheric boundary layer, between 2 km and 13 km of altitude, in the vicinity of the City of Buenos Aires. The CO values were slightly increased without presenting significant risk values for human health. The importance of conducting this type of study is to show that in such aerosols transport events, particles can be transported for hundreds of kilometers from their origin and affect the climate, air quality, and visibility of other areas very distant from the emission source. Satellite measurements, in combination with sun photometers and lidar systems measurements, have allowed an essential synergy for the detection and spatial–temporal monitoring of the smoke columns that, generated thousands of kilometers away, arrived in Argentine territory. Together, these measurements help understand wildfires' environmental impacts in short and long time series, as they provide relevant data for climate and particle dispersion models [51].

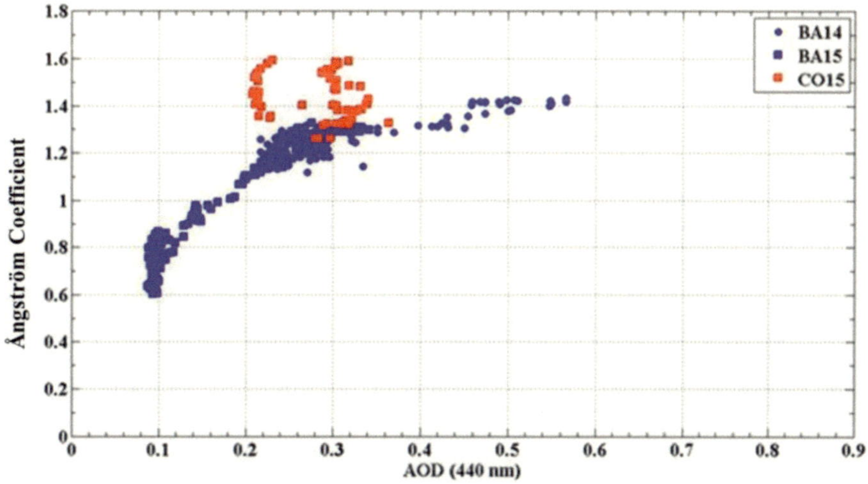

Figure 20.
Ångström coefficient versus AOD (440 nm) for 14 and 15 November 2019 (BA: CEILAP-BA station (blue); CO: Pilar station (red)).

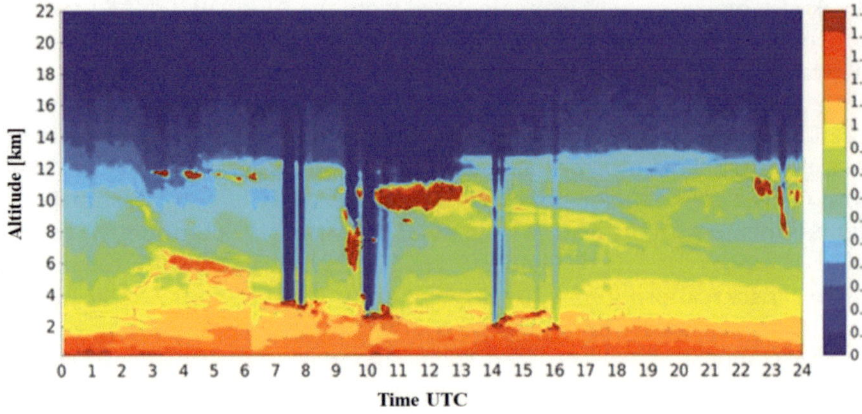

Figure 21.
Spatial–temporal evolution of the normalized aerosols backscattering at 532 nm for 14 November 2019.

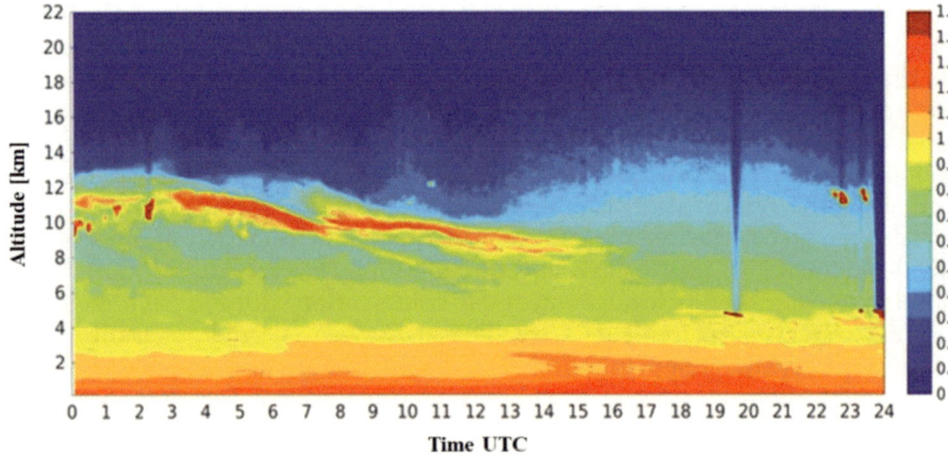

Figure 22.
Spatial–temporal evolution of the normalized aerosols backscattering at 532 nm for 15 November 2019.

4. Dust transport

The long-range transport of desert dust has an essential impact on the atmospheric radiative balance, with a global impact on climate, air quality, and human health [52, 53]. It is estimated that between 1 and 2 billion tons of dust are transported from Sahara each year through the atmosphere [54, 55].

Knowledge of the transport mechanisms of dust aerosols from the Sahara Desert to the South American continent can help understand its impact on the balance of nutrients in the Amazon basin [56, 57] and its nutrients deposition in the equatorial Atlantic [58]. Dust transport from the Sahara over the Atlantic Ocean can reach distances greater than 5000 km from its origin and is more intense during December, January and February at 5 N, and during June, July, and August at 20 N latitude. During the boreal winter, desert aerosols are transported across the Atlantic towards the northeast coast of South America, and in the summer, they can reach the Caribbean Sea [59].

The most favorable periods to observe the presence of Saharan dust with a lidar in the atmosphere of the city of Natal (located about 5 S) are in December, January, and February. In this period, Oliveira et al. [60] found the mineral dust predominance occurred during February, followed by December, January, and March in 2018. Research with lidar on aerosols in the Natal atmosphere focuses on the long-range transport of mineral dust and also smoke from biomass burning from the African continent, occurring under the strong influence of the trade winds, during austral summer, when the Intertropical Convergence Zone (ITCZ) is positioned further to the south [61, 62]. In a case study on 09 February 2018 (Reference [60]), Oliveira et al. show that aerosol plumes were identified below 4000 m altitude.

Landulfo et al. were the first to identify an aerosol plume with a lidar, at 3000 m altitude in Natal's atmosphere, on 01 June 2016 [63]. Direct measurement of the Sun and radiances of the sky with a CIMEL solar photometer may also be used to identify the measured dust aerosol, as in Ref. [60]. Similar aerosol identification studies were carried out [57, 61, 62, 64, 65], showing that African dust's transatlantic transport reaches the Caribbean. Episodes of mineral dust particles transported from Africa to the Amazon between January and April were also observed [66].

The DUSTER Lidar system started operating in 2016 with operational data collection campaigns called MOnitoring aerosol LOng-range Transportation OVer Natal (MOLOTOV) Zero (March 2016 to July 2016.), MOLOTOV I (December 2016 to mid-February 2017), MOLOTOV II (January and February 2018), and APEL (Assessment of Atmospheric Optical Properties During Biomass Burning Events and Long-Range Transport of Desert Dust) (November and December 2017), at the Federal University of Rio Grande do Norte (UFRN), in Natal.

Also, during the campaigns MOLOTOV Zero and MOLOTOV I, eighteen cirrus cloud profiles were analyzed (7 during MOLOTOV Zero and 11 from MOLOTOV I). Part of the research was conducted to calibrate the DUSTER system, applying the $\Delta 90°$ method [67]. The main physical parameters of cirrus clouds obtained from lidar data were their thickness, base height, top height, and linear particle depolarization (δp) [68, 69]. The determination of parameters such as base height and top height was calculated from the analysis of the depolarization profiles obtained for each data interval. Data from radiosondes launched by the Natal airport (about 7 km from the DUSTER system) were used to obtain the average cirrus temperature. The average depolarization value for cirrus clouds during the MOLOTOV Zero campaign was $\delta p(\text{cirrus}) = 0.43 \pm 0.15$ with an average base height of 14.23 km and an average top height of 15.53 km [70]. The average temperature of cirrus during this campaign was −69.23°C. The average thickness of the cirrus found was 1.30 km. For the MOLOTOV I campaign, the average value was $\delta p(\text{cirrus}) = 0.49 \pm 0.13$ with the average base height equal to 12.96 km and the average top height equal to 14.44 km. The average thickness found was 1.48 km. The average cirrus temperature found for this campaign was −62.06°C [70].

5. Cloud studies

Clouds play an important role in Earth's radiation budget as they reflect incoming solar radiation and absorb outgoing thermal radiation emitted from the surface, the atmosphere, and other clouds [71]. In tropical regions, Cirrus clouds are omnipresent [72] and hence affect the radiation balance significantly because of their sizable horizontal extent (hundreds to thousands of km) and long lifetime (hours to days). Cirrus optical properties, altitude, vertical and horizontal coverage control, radiative forcing, and detailed measurements are of absolute value [73, 74].

Studies reporting cirrus properties over tropical South America used to be scarce, with most studies based on in-frequent satellite observations [72, 75]. For obtaining detailed geometrical and microphysical properties of Cirrus clouds, especially the sub-visible type, ground-based lidars are indispensable. For this reason, many lidar groups around the world have used such measurements to obtain and report these characteristics in mid-latitudes [74] and tropical regions [76]. A similar effort is being made in South America, mostly under the auspices of LALINET. Here we review studies from five LALINET stations (Cuba, Manaus, Natal, São Paulo, and Punta Arenas), and we also discuss results from Reference [77], who report on measurements in Buenos Aires. It is compiled a set of statistics for optical and geometrical properties of cirrus occurring in this side of the world, summarized in **Table 2**.

The first studies were performed in Cuba by the lidar group in Camagüey using 6-years (1993 to 1998) elastic lidar data [78]. The lidar measurements were performed for detecting aerosols in the stratosphere, and hence were conducted mostly on clear nights to the naked eye, thus introducing a bias in the cloud measurements towards subvisual cirrus clouds. Indeed, from 131 clouds measured, only 8% were thick (COD >0.3), 67% were thin (0.03 < COD <0.3) and 25% were sub-visual (COD <0.03). Sub-visual and thin cirrus have an average cloud top and base at 14.1 and 11.6 km, respectively, with thick clouds occurring at slightly lower altitudes. The authors estimated the respective optical depths to be 0.50 ± 0.27, 0.07 ± 0.05, and 0.02 ± 0.01, but these are an upper limit as they were calculated assuming a lidar ratio (LR) of 10 sr, which we now know is too low.

In a follow-up study, Barja and Antuña performed radiative transfer simulations (GFDL model, Freidenreich and Ramaswamy, 1999) of the impact of cirrus clouds on solar radiation (SW) [83]. They have found that the daily mean value of SW cirrus radiative forcing (CRF_{SW}) has an average value of −9.1 W m^{-2} at the top of the atmosphere (TOA) and −5.6 W m^{-2} at the surface (SFC). There is a linear relation between CRF_{SW} and COD, with a slope of - 30 W m^{-2}/COD at TOA. The local radiative heating effect where the cirrus is found ranged from 0.35 to 1.24 K day^{-1}, with an average of 0.63 K day^{-1}. These results were the first to show that tropical cirrus radiative impact on Earth's energy balance is essential.

In the Amazon region, Gouveia et al. used continuous measurements (July 2011 to June 2012) at the Manaus station [1] to retrieve the optical and geometrical properties of cirrus clouds during day and night [79]. The cirrus frequency of occurrence was about 88% during the wet season and not lower than 50% during the dry season, with a mean column optical depth of 0.25 ± 0.46. The cloud top and base heights, as well as cloud thickness and cloud optical depth, were, respectively, 14.3 ± 1.9 (std) km, 12.9 ± 2.2 km, 1.4 ± 1.1 km, and 0.25 ± 0.46, similar to the values reported in Cuba. These clouds have a significant radiative impact with such a high frequency of occurrence and altitude over the dark-pristine Amazon forest (albedo about 0.12).

Gouveia then used these measured optical depths and vertical profiles of the cirrus extinction coefficient [84], at 5-min and 200 m resolutions, and calculated the cirrus radiative forcing (CRF) with libRadtran [85]. Cirrus in the Amazon region produced a net CRF at TOA and SFC of +15.3 and − 3.7 W m^{-2}, respectively, much more intense than predicted for 3 European sites (+0.9, +1, and + 1.7 W m^{-2} at TOA) [74, 86]. Optically thicker cirrus, in general, have more prominent net CRF, with instantaneous CRF that could reach extremes up (down) to +140 (−65) W m^{-2} for the night (day) time [85]. Together, the vertical profiles with total COD >0.3 were responsible for about 72% (62%) of the TOA (BOA) net CRF, which means that a large fraction of the CRF is generated by optically thin cirrus (COD <0.3) that are harder to detect by radars and passive instruments on board of satellites [86]. A definite daily cycle of the optical depth was found and shows a minimum about local noon and a maximum in the late afternoon (~16 h LT), associated with the diurnal precipitation cycle. This results in a mean instantaneous TOA (SFC)

Loc.	Camagüey, Cuba	Manaus, Brazil	Natal, Brazil	São Paulo, Brazil	B. Aires, Argentina	P. Arenas, Chile
Ref.	[78]	[79]	[80]	[81]	[77]	[82]
Lat.	21 N	3 S	6 S	23 S	35 S	53 S
Period	1993–1998	2011–2012	Jan-Feb 2017–2018	Jun-Jul 2007	2001–2005	2016–2018
λ (nm)	532	355	532	532	532	532
Freq.	41%	74%	67%	54%	—	45%
Top	14.1 ± 1.4	14.3 ± 1.9	13.4 ± 1.6	12.4 ± 0.6	11.8	10.8 ± 2.2
Base	11.6 ± 1.5	12.9 ± 2.2	11.5 ± 1.9	10.2 ± 0.7	9.6	9.0 ± 2.4
COD	0.50 ± 0.27	0.35 ± 0.55	—	0.27 ± 0.22	—	0.25 ± 0.32
LR (sr)	10 (fixed)	23.6 ± 8.1	—	26 ± 12	—	—

Table 2.
Cirrus clouds' properties were measured by different LALINET stations, from north to south, during the last 15 years. Columns indicate latitude (deg) of the station, period of study and wavelength, frequency of occurrence, cirrus layer average top and base altitudes, column cloud optical depth (COD), and lidar ratio (LR).

net CRF ranging from +1.7 (−23) W m^{-2} in the afternoon to +47 (+3.1) W m^{-2} at night [86]. The cirrus clouds produced an average in-cloud heating rate ranging between −1 K day^{-1} to +2 K day^{-1} vertical profile from 8 to 18 km (in-cloud), but with instantaneous values that can reach values higher than 10 K day^{-1} for portions of the cloud with high ice water content [86].

The DUSTER Lidar station at Natal, Brazil, is the most recent addition to the network. Cirrus measurements started only recently, and Santos reports on measurements during January–February (pre-rainy season) of 2017 and 2018 [80]. A total number of 35 clouds were studied and showed cirrus up to 16 km. These clouds are lower than those in Manaus and were never observed above the tropopause at 17–18 km. The reason is the less vigorous convection in this coastal site, which resembles an oceanic precipitation regime. The frequency of occurrence and average cloud top altitude was 74% (57%) and 13.9 km (12.3 km) in 2017 (2018), respectively. In situ data obtained by radiosondes (9.5 km away) for selected case studies showed an increase of the relative humidity in the layer where the lidar identified the cirrus clouds, from around 10–20% below/above the cloud to around 40–55% in the cloud altitude region.

Cirrus clouds over São Paulo, in the subtropics of South America, were studied by Larroza [81]. She analyzed 34 days, from June to July 2007, using the methodology described in Ref. [87]. The cirrus frequency of occurrence was 54%. The vertical distribution of cloud tops showed peaks at 9.6, 10.6, 12.3, and 13.9 km, with an average of 12.4 km. These cirri were optically thinner (0.27) and occurred at lower altitudes (cloud top 12.4 km) than their tropical counterpart but had a similar lidar ratio of about 26 sr. The clouds observed were either produced by the passage of cold fronts or transported from the tropics or mid-latitudes.

Going into the mid-latitudes, Lakkis et al. used data from a lidar system in Buenos Aires, Argentina, that was not part of LALINET [77]. They studied 60 diurnal cirrus clouds from 2001 to 2005. Unlike the tropics and sub-tropics, cirrus tops were only found very close to the tropopause (~380 m), with cloud tops at 11.8 ± 0.86 and bases at 9.6 ± 0.9 km. Unfortunately, the low statistics did not allow the calculation of a frequency of occurrence, nor did the authors report values of optical depth or lidar ratios. The southernmost LALINET station is at Punta Arenas, Chile (53°S, 71°W), a sub-Antarctic region. Lidar cirrus measurements there began in October 2016 and continue to the present. A preliminary result of cirrus clouds' geometric characteristics in the region, over two years (from 2016 to 2018), shows that the cirrus' mean base height is 9.0 ± 2.4 km and the mean top height is 10.8 ± 2.2 km. In the same site in November 2018, the Leipzig Aerosol and Cloud Remote Observations System (LACROS) [88] was deployed by the Leibniz Institute for Tropospheric Research (TROPOS) in collaboration with the University of Magallanes (UMAG) for the field experiment DACAPO - PESO (Dynamics, Aerosol, Cloud And Precipitation Observations in the Pristine Environment of the Southern Ocean). Cirrus clouds measurements are being performed with a Raman polarization lidar that will allow the calculation of LR and COD at multiple wavelengths, which will be reported in a future study.

There has been a great effort to study cirrus clouds in South America, with measurements taking place from 1993 to today, and from 21 N to 53 S. **Table 3** summarizes the main characteristics, and we can see how the cirrus altitude changes from the Tropical (high, 14.3 km) to the sub-Antarctic regions (low 10.8 km). The frequency of occurrence also becomes smaller, reducing from 74% in Manaus to 45% in Punta Arenas. Deep convection, prevalent in the tropics and sub-tropics, pushes the tropopause upward and creates optically and physically thick cirrus clouds from the anvil's detrainment. From the subtropics towards the polar regions, convection is limited by the lack of surface heating, and the mixing of tropospheric air depends on

		CRF_{SW}	CRF_{LW}	CRF_{NET}
Barja and Antuña Camagüey, Cuba	**SFC**	−5.6	—	—
	TOA	−9.1	—	—
Gouveia Manaus, Brazil	**SFC**	−5.0 ± 0.4	1.25 ± 0.04	−3.7 ± 0.2
	TOA	−5.7 ± 0.5	21.0 ± 0.6	15.3 ± 0.4

Table 3.
Cirrus Clouds radiative forcing (W m^{-2}) over Cuba [83] and the central Amazon forest [79] calculated by radiative transfer simulations based on ground-based lidar measurements.

uplift by frontal systems. This does not reach high altitudes, and the primary cirrus production mechanism will be the large-scale lifting of water vapor, rendering physically and optically thinner clouds. Lidars can directly observe the optical depth, and **Table 3** shows that it also becomes smaller, decreasing from 0.35 in Manaus to 0.25 in Punta Arenas. Unfortunately, there are not enough LR estimations to allow for a comparison. They were calculated only for Manaus and São Paulo, and the results are in close agreement, indicating similar crystal habits and formation mechanisms, as expected. Similar relation of the cirrus features and latitude was first reported by Cordoba-Jabonero et al. using data from the LALINET subtropical station of São Paulo and ground-based lidar located in Belgrano Antarctic Station [88].

Another aspect of this LALINET effort is the diversity in the methods and the timespan of the different studies, limiting our ability for a more in-depth comparison. The combination of the Klett and transmittance methods, as described by Larroza [87] and Gouveia et al. [79], can be applied to elastic data from any of the LALINET lidars, providing COD and LR with high temporal resolution (e.g., 5-min), during both day and night. The use of radiative transfer models to calculate the cirrus' radiative impact could also be performed for all stations doing cirrus measurements. It would be essential to homogenize cirrus clouds' analysis throughout the network groups, even by sharing analysis algorithms. Moreover, it should be emphasized that these critical analyses could be automatized and performed unattended on a unique central server. These facts highlight the importance of establishing systematic data sharing in the context of LALINET and GALION.

6. Conclusions

Tropospheric lidars can provide impressive results: aerosol studies related to volcanic eruptions, tropospheric systems, biomass burning, and dust transport (both into and out of the continent) were conducted over the past 15 years. Pollutant dispersion studies in large cities of South America employing PBL dynamics, conducted in different sites, and with specific conditions, greatly aid local air quality authorities. More recently, studies on clouds and their impact on radiative transfer have been carried out. Given the geographical location of the lidar sites, essential comparisons can be carried on along a wide longitudinal interval, and tropospheric-stratospheric circulation experiments can be performed, as well as their climatological interpretation.

Acknowledgements

The authors are thankful to the Brazilian Agencies National Council for Scientific and Technological Development (CNPq), Coordination for the

Improvement of Higher Education Personnel (CAPES), São Paulo Research Foundation (FAPESP), Brazilian Agricultural Research Corporation (EMBRAPA), and National Institute of Amazonian Research (INPA) LBA Central Office in Manaus. The authors also thank the NASA/AERONET teams, Japan International Cooperation Agency (JICA), the Argentine Agencies National Scientific and Technical Research Council (CONICET), National Agency for the Promotion of Research, Technological Development and Innovation (ANPCyT), the Argentine National Defense University (UNDEF), UNDEFI and PID-UTN Projects, the Ministry of Defense of Argentina, and the French National Centre for Scientific Research (CNRS). Also, to all NASA's technical personnel, the Argentine Institute of Scientific and Technical Research for Defense (CITEDEF), and the Argentine National Meteorological Service (SMN), who have kept the solar photometers in operation, and especially to Raúl D'Elia. The authors wish to acknowledge the entire NASA CALIPSO and MODIS (AQUA/TERRA) teams, the NOAA Air Resources Laboratory, for providing the HYSPLIT transport and dispersion model and the READY website, ESA/EOM projects teams, the Suomi NPP (National Polar-orbiting Partnership) Mission teams, and the Sentinel 5-P TROPOMI team. The authors also acknowledge the financial support from CIBioFi, the Colombian Science, Technology, and Innovation Fund-General Royalties System (Fondo CTeI-Sistema General de Regalías), and Gobernación del Valle del Cauca. The authors acknowledge the China-Brazil Joint Laboratory for Space Weather (CBJLSW) for Supporting this Book Chapter. Vania F. Andrioli would like to thank the CBJLSW and the National Space Science Center (NSSC) of the Chinese Academy of Sciences (CAS) for supporting her postdoctoral fellowship. The authors from the Universidad de Magallanes would like to acknowledge the financial support of the Japan Science and Technology Agency (JST) /Japan International Cooperation Agency (JICA), the Science and Technology Research Association for Sustainable Development (SATREPS) through the SAVERNet project; and the Program FONDECYT of the Chilean National Agency for Research and Development (ANID) through Project FONDECYT 11181335.

Conflict of interest

The authors declare no conflict of interest.

Author details

Eduardo Landulfo[1*], Alexandre Cacheffo[1,2], Alexandre Calzavara Yoshida[1,2], Antonio Arleques Gomes[1], Fábio Juliano da Silva Lopes[1,3], Gregori de Arruda Moreira[1,4,5], Jonatan João da Silva[1,6], Vania Andrioli[7,8], Alexandre Pimenta[7], Chi Wang[9], Jiyao Xu[9], Maria Paulete Pereira Martins[7], Paulo Batista[7], Henrique de Melo Jorge Barbosa[10,11], Diego Alves Gouveia[10,12], Boris Barja González[13], Felix Zamorano[13], Eduardo Quel[14], Clodomyra Pereira[15,16], Elian Wolfram[17,18], Facundo Ismael Casasola[15,16,19], Facundo Orte[18], Jacobo Omar Salvador[18], Juan Vicente Pallotta[18], Lidia Ana Otero[14,19], Maria Prieto[15,16], Pablo Roberto Ristori[14], Silvina Brusca[14], John Henry Reina Estupiñan[20,21], Estiven Sanchez Barrera[20], Juan Carlos Antuña-Marrero[22], Ricardo Forno[23], Marcos Andrade[23], Judith Johanna Hoelzemann[24], Anderson Guimarães Guedes[25], Cristina Tobler Sousa[24], Daniel Camilo Fortunato dos Santos Oliveira[26,27], Ediclê de Souza Fernandes Duarte[27], Marcos Paulo Araújo da Silva[26,27] and Renata Sammara da Silva Santos[1,24]

1 Center for Lasers and Applications (CELAP), Institute of Energy and Nuclear Research (IPEN) Sao Paolo, Brazil

2 Institute of Exact and Natural Sciences of Pontal (ICENP), Federal University of Uberlândia (UFU), Ituiutaba, Brazil

3 Institute of Environmental, Chemical and Pharmaceutical Sciences (ICAQF), Federal University of São Paulo (UNIFESP), Diadema, Brazil

4 Federal Institute of São Paulo (IFSP), Campus Registro, Sao Paolo, Brazil

5 Institute of Astronomy, Geophysics and Atmospheric Sciences (IAG), University of São Paulo (USP), Sao Paolo, Brazil

6 Center for Exact Sciences and Technologies (CCET), Federal University of Western Bahia (UFOB), Barreiras, Brazil

7 National Institute for Space Research (INPE), São José dos Campos, Brazil

8 China-Brazil Joint Laboratory for Space Weather (NSSC/INPE), São José dos Campos, Brazil

9 State Key Laboratory of Space Weather (SKSW), National Space Science Center (NSSC), Chinese Academy of Sciences (CAS), Beijing, China

10 Physics Institute, University of São Paulo (USP), Sao Paolo, Brazil

11 Physics Department, University of Maryland Baltimore County (UMBC), Baltimore, USA

12 Royal Netherlands Meteorological Institute (KNMI), De Bilt, The Netherlands

13 Department of Mathematics and Physics, University of Magallanes (UMAG), Punta Arenas, Chile

14 Institute of Scientific and Technical Research for Defense (CITEDEF) - UNIDEF (MINDEF – CONICET), Buenos Aires, Argentina

15 Military Geographical Service, National Geographic Institute (IGN), Buenos Aires, Argentina

16 General Directorate of Research and Development of the Argentine Army (DIGID), Buenos Aires, Argentina

17 Argentine National Weather Service (SMN), Buenos Aires, Argentina

18 Laser and Applications Research Center (CEILAP), UNIDEF (MINDEF–CONICET), Buenos Aires, Argentina

19 Argentine National Defense University (UNDEF), Army Engineering Faculty (FIE), Buenos Aires, Argentina

20 Centre for Bioinformatics and Photonics (CIBioFi), Universidad del Valle (UniValle), Cali, Colombia

21 Physics Department, Universidad del Valle (UniValle), Cali, Colombia

22 Department of Theoretical, Atomic and Optical Physics, University of Valladolid (UVA), Valladolid, Spain

23 Department of Physics, Major University of San Andrés (UMSA), La Paz, Bolivia

24 Department of Atmospheric and Climate Sciences (DCAC), Federal University of Rio Grande do Norte (UFRN), Natal, Brazil

25 School of Science and Technology (ECT), Federal University of Rio Grande do Norte (UFRN), Natal, Brazil

26 Polytechnic University of Catalonia (UPC), Barcelona, Spain

27 Graduate Program in Climate Sciences, Federal University of Rio Grande do Norte (UFRN), Natal, Brazil

*Address all correspondence to: landulfo@gmail.com; elandulf@ipen.br

References

[1] Barbosa, H. M. J., Barja, B., Pauliquevis, T., Gouveia, D. A., Artaxo, P., Cirino, G. G., Santos, R. M. N., Oliveira A. B., 2014: A permanent Raman lidar station in the Amazon: Description, characterization and first results. Atmos. Meas. Tech., v. 7, Issue 6, 1745-1762. https://doi.org/10.5194/amt-7-1745-2014.

[2] Landulfo, E., Papayannis, A., Artaxo, P., Castanho, A. D. A., de Freitas, A. Z., Souza, R. F., Vieira Junior, N. D., Jorge, M. P., Sánchez-Ccoyllo, O. R., Moreira, D. S., 2003: Synergetic measurements of aerosols over São Paulo, Brazil using LIDAR, sunphotometer and satellite data during the dry season, Atmos. Chem. Phys., v. 3, Issue 5, 1523-1539. https://doi.org/10.5194/acp-3-1523-2003.

[3] Landulfo, E., Matos, C. A., Torres, A. S., Sawamura, P., Uehara, S. T., 2007: Air quality assessment using a multi-instrument approach and air quality indexing in an urban area. Atmos. Res., v. 85, Issue 1, 98-111. https://doi.org/10.1016/j.atmosres.2006.11.008.

[4] Landulfo, E., Jorge, M. P. M. P., Held, G., Guardani, R., Steffens, J., Pinto, S. A. F., et al., 2010: Lidar observation campaign of sugar cane fires and industrial emissions in the State of São Paulo, Brazil. Proceedings of SPIE - The International Society for Optical Engineering. 7832. Lidar Technologies, Techniques, and Measurements for Atmospheric Remote Sensing VI. https://doi.org/10.1117/12.866078.

[5] Landulfo, E., Costa, R. F., Rodrigues, P., Lopes, F. J. S., 2004: Raman lidar characterization using a reference lamp. Proc. SPIE, Lidar Technologies, Techniques, and Measurements for Atmospheric Remote Sensing X, v. 9246. https://doi.org/10.1117/12.2057376.

[6] Oliveira, A. M., Sousa, C. T., Oliveira, N. P. M., Melo, A. K. S., Lopes, F. J. S., Landulfo, E., Elbern, H., Hoelzemann, J. J., 2019: Analysis of Atmospheric Aerosol Optical Properties in the Northeast Brazilian Atmosphere with Remote Sensing Data from MODIS and CALIOP/CALIPSO Satellites, AERONET Photometers and a Ground-Based Lidar. Atmosphere, v. 10, Issue 10, 594-615. https://doi.org/10.3390/atmos10100594.

[7] Guedes, A. G., Landulfo, E., Montilla-Rosero, E., Lopes, F. J. S., Hoelzemann, J. J., Fernandez, J. H., et al., 2018: Detection of Saharan mineral dust aerosol transport over Brazilian northeast through a depolarization lidar. EPJ Web of Conferences, v. 176, 05036. https://doi.org/10.1051/epjconf/201817605036.

[8] The European Aerosol Research Lidar Network: EARLINET. Available from http://www.earlinet.org. Accessed 30 October 2020.

[9] The National Institute for Research and Development for Optoelectronics (INOE), Romania. Remote Sensing. International Projects. About APEL. Available from: http://environment.inoe.ro/article/237/about-apel. Accessed 30 October 2020.

[10] Lopes, F. J. S., Moreira, G. A., Guerrero-Rascado, J. L., Silva, J. A., Gouveia, D. A., Gomes, A., Belegante, L., Freudenthaler, V., et al., 2018: Improving the Instrument and Analysis Capabilities of the São Paulo LALINET Lidar Station in the Framework of the APEL Project. In: 1st European Lidar Conference ELC2018, Thessaloniki, Greece. Available from: http://www.fap.if.usp.br/~hbarbosa/uploads/Site/Publications/Fabio_elc2018.pdf. Accessed 30 October 2020.

[11] Nicolae, D., Guerrero-Rascado, J. L., Belegante, L., Freudenthaler, V., Landulfo, E., Lopes, F. J. S.,

Alados-Arboledas, L., et al., 2018: APEL: Assessment of atmospheric optical Properties during biomass burning Events and Long-range transport of desert dust. Alcantara Study Executive Summary. ESA Contract N. 4000117289. Available from: https://nebula.esa.int/sites/default/files/neb_study/1317/C4000117289ExS.pdf. Accessed 30 October 2020.

[12] D'Amico, G., Amodeo, A., Baars, H., Binietoglou, I., Freudenthaler, V., Mattis, I., Wandinger, U., Pappalardo, G., 2015: EARLINET Single Calculus Chain – overview on methodology and strategy. Atmos. Meas. Tech., v. 8, Issue 11, 4891-4916. https://doi.org/10.5194/amt-8-4891-2015.

[13] D'Amico, G., Amodeo, A., Mattis, I., Freudenthaler, V., Pappalardo, G., 2016: EARLINET Single Calculus Chain – technical – Part 1: Pre-processing of raw lidar data. Atmos. Meas. Tech., v. 9, Issue 2, 491-507. https://doi.org/10.5194/amt-9-491-2016.

[14] Mattis, I., D'Amico, G., Baars, H., Amodeo, A., Madonna, F., Iarlori M., 2016: EARLINET Single Calculus Chain – technical – Part 2: Calculation of optical products. Atmos. Meas. Tech., v. 9, Issue 7, 3009-3029. https://doi.org/10.5194/amt-9-3009-2016.

[15] Wiedensohler, A., Andrade, M., Weinhold, K., Müller, T., Birmili, W., Velarde, et al., 2018: Black carbon emission and transport mechanisms to the free troposphere at the La Paz/El Alto (Bolivia) metropolitan area based on the Day of Census (2012), Atmos. Environ., v. 194, 158-169. https://doi.org/10.1016/j.atmosenv.2018.09.032.

[16] Stull, R. B., 1988: An Introduction to Boundary Layer Meteorology. Atmospheric and Oceanographic Sciences Library, v. 13. Springer, Dordrecht. ISBN: 978-90-277-2769-5. https://doi.org/10.1007/978-94-009-3027-8.

[17] Holzworth, C. G., 1964: Estimates of mean maximum mixing depths in the contiguous United States. Month. Weath. Rev., v. 92, Issue 5, 235-242. https://doi.org/10.1175/1520-0493(1964)092<0235:EOMMMD>2.3.CO;2.

[18] Zeng X., Brunke, M. A., Zhou, M., Fairall, C., Bond, N. A., Lenschow, D. H., 2004: Marine atmospheric boundary layer height over the eastern Pacific: data analysis and model evaluation. J. Climate, v. 17, Issue 21, 4159-4170. https://doi.org/10.1175/JCLI3190.1.

[19] Seidel, D. J., Ao, C. O., Li, K., 2010: Estimating climatological planetary boundary layer heights from radiosonde observations: Comparison of methods and uncertainty analysis. J. Geophys. Res., v. 115, Issue D16, D16113. https://doi.org/10.1029/2009JD013680.

[20] Emeis, S., Schäfer, K., Münkel, C., 2008: Surface-based remote sensing of the mixing-layer height - a review. Meteorologische Zeitschrift, v. 17, Issue 5, 621-630. https://doi.org/10.1127/0941-2948/2008/0312.

[21] Moreira, G. A., Landulfo, E., Peres, L. V., Mariano, G., Borayou, R., 2013: Obtenção da altura da Camada Limite Planetária a partir do Método das Imagens (In Portuguese). Rev. Ciência e Natura. Santa Maria, Brazil. Special Edition, 335 - 338. ISSN-e: 2179-460X. https://doi.org/10.5902/2179460X11645.

[22] Moreira, G. A., Lopes, F. J. S., Held, G., Albuquerque, T. T. A., Reis Junior, N. C., Borayou, R., Landulfo, E., 2013: Automatic methods to detect the top of the Atmospheric Boundary Layer. Proc. Lidar Technologies, Techniques, and Measurements for Atmospheric Remote Sensing IX. 88940T. https://doi.org/10.1117/12.2028750.

[23] Nisperuza, D. J., 2014: Lidar measurements and wavelet covariance transform method to estimate the

atmospheric boundary layer heights in Medellín, Colombia. Ópt. Pura y Aplicada, v. 47, Issue 2, 123-130. https://doi.org/10.7149/opa.47.2.123.

[24] Moreira, G. A., Marques, M. T. A., Nakaema, W., Moreira, A. C. de C. A., Landulfo, E., 2015: Planetary Boundary Layer height estimation from Doppler wind lidar measurements, radiosonde and HYSPLIT model comparison. Óptica Pura e Aplicada, v. 48, Issue 3, 179-183. ISSN-e 2171-8814. https://doi.org/10.7149/OPA.48.3.179.

[25] Moreira, G. A., Marques, M. T. A., Nakaema, W., Moreira, A. C. de C. A., Landulfo, E., 2015: Detecting the planetary boundary layer height from low-level jet with Doppler lidar measurements. Proc. SPIE 9645: Lidar Technologies, Techniques, and Measurements for Atmospheric Remote Sensing XI. 96450F. https://doi.org/10.1117/12.2195278.

[26] Salvador, N., Loriato, A. G., Santiago, A., Albuquerque, T. T. A., Reis Jr., N. C., Santos, J. M., et al., 2016: Study of the Thermal Internal Boundary Layer in Sea Breeze Conditions Using Different Parameterizations: Application of the WRF Model in the Greater Vitória Region. Rev. Bras. Meteorol., v. 31, Issue 4, 593 - 609. http://doi.org/10.1590/0102-7786312314b20150093.

[27] Céspedes, J., Melo-Luna, C. A., Reina, J. H., 2018: A first measurement of the planetary boundary layer top in Santiago de Cali - Colombia: elastic LiDAR application. Proc. Remote Sensing of Clouds and the Atmosphere XXIII. 107860L. https://doi.org/10.1117/12.2503534.

[28] Marques, M. T . A., Moreira, G. A., Piñero, M., Oliveira, A. P., Landulfo, E., 2018: Estimating the planetary boundary layer height from radiosonde and doppler lidar measurements in the city of São Paulo – Brazil, EPJ Web Conf., v. 176, 06015. https://doi.org/10.1051/epjconf/201817606015.

[29] Herrera-Mejía, L., Hoyos, C. D., 2019: Characterization of the atmospheric boundary layer in a narrow tropical valley using remote-sensing and radiosonde observations and the WRF model: the Aburrá Valley case-study. Q. J. R. Meteorol. Soc., v. 145, Issue 723, 2641 – 2665. https://doi.org/10.1002/qj.3583.

[30] Moreira, G. A., Lopes, F. J. S., Guerrero-Rascado, J. L., da Silva, J. J., Gomes, A., A., Landulfo, E., Alados-Arboledas, L., 2019: Analyzing the atmospheric boundary layer using high-order moments obtained from multiwavelength lidar data: impact of wavelength choice. Atmos. Meas. Tech., v. 12, Issue 8, 4261-4276. https://doi.org/10.5194/amt-12-4261-2019.

[31] Vivas, D. R., Sánchez, E., Reina, J. H., 2020: Deep learning the atmospheric boundary layer height. arXiv. preprint. arXiv: 2004.04353v1 [physics.ao-ph]. Available from: https://arxiv.org/pdf/2004.04353.pdf. Accessed 15 October 2020.

[32] Hayden, K., Anlauf, K., Hoff, R., Strapp, J., Bottenheim, J., Wiebe, H., et al., 1997: The vertical chemical and meteorological structure of the boundary layer in the Lower Fraser Valley during Pacific'93. Atmospheric Environment, v. 31, Issue 14, 2089-2105. https://doi.org/10.1016/S1352-2310(96)00300-7.

[33] Menut, L., Flamant, C., Pelon, J., Flamant, P. H., 1999: Urban boundary-layer height determination from lidar measurements over the Paris area. Applied Optics, v. 38, Issue 6, 945-954. https://doi.org/10.1364/AO.38.000945.

[34] Brooks, I. M., 2003: Finding boundary layer top: Application of a wavelet covariance transform to lidar backscatter profiles. Journal

of Atmospheric and Oceanic Technology, v. 20, Issue 8, 1092-1105. https://doi.org/10.1175/1520-0426(2003)020<1092:FBLTAO>2.0.CO;2.

[35] D'Almeida, G. A., Koepke, P., Shettle. E. P., 1991: Atmospheric aerosols: Global climatology and radiative characteristics. A. Deepak Publishing, Hampton, Virginia. ISBN-10: 0937194220.

[36] Australian Government. Bureau of Meteorology, 2019: Special Climate Statement 70 update – drought conditions in Australia and impact on water resources in the Murray–Darling Basin. Published 29 November 2019. Available from: http://www.bom.gov.au/climate/current/statements/scs70b.pdf. Accessed 20 October 2020.

[37] Randerson, J. T., Chen, Y., van der Werf, G. R., Rogers, B. M., Morton, D. C., 2012: Global burned area and biomass burning emissions from small fires. J. Geophys. Res., v. 117, G04012. https://doi.org/10.1029/2012JG002128.

[38] Suomi NPP (National Polar-orbiting Partnership) Mission. Available from: https://earth.esa.int/web/eoportal/satellite-missions/s/suomi-npp. The ESA Earth Observation Portal (eoPortal). Accessed 20 October 2020.

[39] OMPS-NPP L2 NM Aerosol Index swath orbital V2. Available from: https://disc.gsfc.nasa.gov/datasets/OMPS_NPP_NMMIEAI_L2_2/summary. Accessed 20 October 2020.

[40] Moderate Resolution Imaging Spectroradiometer (MODIS). Available from: https://modis.gsfc.nasa.gov. Accessed 20 October 2020.

[41] Levy, R. C., Mattoo, S., Munchak, L. A., Remer, L. A., Sayer, A. M., Patadia, F., Hsu, N. C., 2013: The Collection 6 MODIS aerosol products over land and ocean. Atmospheric Measurement Techniques, v. 6, 2989-3034. https://doi.org/10.5194/amt-6-2989-2013.

[42] Apituley, A., Pedergnana, M., Sneep, M., Pepijn Veefkind, J., Loyola, D., Landgraf, J., Borsdorff, T., 2018: Sentinel-5 precursor/TROPOMI Level 2 Product User Manual Carbon Monoxide. Royal Netherlands Meteorological Institute (KNMI). Document Number: SRON-S5P-LEV2-MA-002. Available from: https://sentinel.esa.int/documents/247904/2474726/Sentinel-5P-Level-2-Product-User-Manual-Carbon-Monoxide. Accessed 20 October 2020.

[43] Torres, O., Bhartia, P. K., Herman, J. R., Ahmad, Z., Gleason, J., 1998: Derivation of aerosol properties from satellite measurements of backscattered ultraviolet radiation: Theoretical basis. Journal Geophys. Res., v. 103, Issue D14, 17099 – 17110. https://doi.org/10.1029/98JD00900.

[44] Holben, B. N., Eck, T. F., Slutsker, I., Tanré, D., Buis, J. P., Setzer, A., et al., 1998: AERONET - A federated instrument network and data achieve for aerosol characterization. Remote Sens. of Environ., v. 66, Issue 1, 1-16. https://doi.org/10.1016/S0034-4257(98)00031-5.

[45] Dubovik, O., King, M., 2000: A flexible inversion algorithm for retrieval of aerosol optical properties from Sun and sky radiance measurements. J. Geophys. Res., v. 105, Issue D16, 20673-20696. https://doi.org/10.1029/2000JD900282.

[46] Otero L., Ristori, P., Holben, B., Quel, E., 2006: Aerosol Optical Thickness at ten AERONET – NASA stations during 2002 (In Spanish). Opt. Pura y Apl. v. 39, Issue 4, 355-364. Sociedad Española de Óptica. Available from: https://www.sedoptica.

es/Menu_Volumenes/Pdfs/228.pdf. Accessed 20 October 2020.

[47] Ristori, P., Otero, L., Jin, Y., Barja, B., Shimizu, A., Barbero, A., Salvador, J., Bali, J. L., Herrera, M., Etala, P., Acquesta, A., Quel, E., Sugimoto, N., Mizuno, A., 2018: Saver.net lidar network in southern South America. The 28th International Laser Radar Conference (ILRC 28). EPJ Web Conf., v. 176, 09011. https://doi.org/10.1051/epjconf/201817609011.

[48] Otero L., Ristori, P., Holben, B. N., Quel, E., 2006: Aerosol optical properties by means of a sunphotometer and lidar system in Buenos Aires, Argentina. Opt. Pura y Apl., Especial III Workshop on Lidar Measurements in Latin America, v. 39, n. 1, 43 – 47. Sociedad Española de Óptica. e-ISSN: 2171-8814. Available from: https://www.sedoptica.es/Menu_Volumenes/Pdfs/183.pdf. Accessed 20 October 2020.

[49] Otero, L., Ristori, P., Pawelko, E., Pallotta, J., Quel, E., 2011: Six-Year Evolution of Multiwavelength Lidar System at CEILAP. Opt. Pura y Apl., V Workshop on Lidar Measurements in Latin America, v. 44, 13 – 18. Sociedad Española de Óptica. ISSN 0030-3917. Available from: https://www.sedoptica.es/Menu_Volumenes/Pdfs/OPA44-1-13.pdf. Accessed 20 October 2020.

[50] Landgraf, J., aan de Brugh, J., Scheepmaker, R., Borsdorff, T., Hu, H., Houweling, S., Butz, A., Aben, I., and Hasekamp, O., 2016: Carbon monoxide total column retrievals from TROPOMI shortwave infrared measurements. Atmos. Meas. Tech., v. 9, 4955-4975. https://doi.org/10.5194/amt-9-4955-2016.

[51] Otero L., Casasola, F., Pereyra, C., Prieto, M., Brusca, S., Ristori, P., 2020: Australian Aerosol Layers Over Argentine Territory During November 2019 (In Spanish). Asociación Física Argentina. Anales AFA, v. 31, n. 1, 1-6. ISSN : 1850-1168. https://doi.org/10.31527/analesafa.2020.31.1.1.

[52] Mahowald, N. M., Ballantine, J. A., Feddema, J., and Ramankutty, N., 2007: Global trends in visibility: implications for dust sources. Atmos. Chem. Phys., v. 7, Issue 12, 3309-3339. https://doi.org/10.5194/acp-7-3309-2007.

[53] Prospero, J. M., Collard, F-X., Molinié, J., Jeannot, A., 2014: Characterizing the annual cycle of African dust transport to the Caribbean Basin and South America and its impact on the environment and air quality. Global Biogeochemical Cycles, v. 28, Issue 7, 757-773. https://doi.org/10.1002/2013GB004802.

[54] Zender, C. S., Miller, R. L. R. L., Tegen, I., 2004: Quantifying mineral dust mass budgets: Terminology, constraints, and current estimates. EOS. Science News by AGU, v, 85, Issue 48, 509-512. https://doi.org/10.1029/2004EO480002.

[55] Mona L, Liu Z, Müller D, Omar A, Papayannis A, Pappalardo G, et al. Lidar Measurements for Desert Dust Characterization: An Overview. Advances in Meteorology, v. 2012, Article ID 356265. In: Special Issue. Desert Dust Properties: Modelling, and Monitoring; 2012 https://doi.org/10.1155/2012/356265

[56] Bristow, C. S., Hudson-Edwards, K. A., Chappell, A., 2010: Fertilizing the Amazon and equatorial Atlantic with West African dust. Geophysical Research Letters: Atmospheric Science, v. 37, Issue 14. https://doi.org/10.1029/2010GL043486.

[57] Ben-Ami, Y., Koren, I., Rudich, Y., Artaxo, P., Martin, S. T., and Andreae, M. O., 2010: Transport of North African dust from the Bodélé depression to the Amazon Basin: a case

study, Atmos. Chem. Phys., v. 10, Issue 16, 7533-7544. https://doi.org/10.5194/acp-10-7533-2010.

[58] Schulz, M., Prospero, J. M., Baker, A. R., Dentener, F., Ickes, L., Liss, P. S., et al., 2012: The atmospheric transport and deposition of mineral dust to the ocean: Implications for research needs. Environ. Sci. Technol., v. 46, Issue 19, 10390-10404. https://doi.org/10.1021/es300073u.

[59] Ben-Ami, Y., Koren, I., Altaratz, O., 2009: Patterns of Saharan dust transport over the Atlantic: Winter vs. Summer, based on CALIPSO first year data. Atmospheric Chemistry and Physics, v. 9, Issue 3, 7867-7865. https://doi.org/10.5194/acpd-9-13177-2009.

[60] Oliveira, D. C. F. S., Montilla-Rosero, E., Lopes, F. J. S., Morais, F. G., Landulfo, E., Hoelzemann, J. H., 2020: Aerosol properties in the atmosphere of Natal/Brazil measured by an AERONET Sun-photometer. Environmental Science and Pollution Research. https://doi.org/10.1007/s11356-020-11373-z.

[61] Talbot, R. W., Andreae, M. O., Berresheim, H., Artaxo, P., Garstang, M., Harriss, R. C., Beechef, K. M., Li, S. M., 1990: Aerosol chemistry during the wet season in central Amazonia: The influence of long-range transport. J. Geophys. Res., v. 95, Issue D10, 16955-16969. https://doi.org/10.1029/JD095iD10p16955.

[62] Swap, R., Garstang, M., Greco, S., Talbot, R., Kållberg, P., 1992: Saharan dust in the Amazon Basin. Tellus B, v. 44, Issue 2, 133-149. https://doi.org/10.1034/j.1600-0889.1992.t01-1-00005.x.

[63] Landulfo, E., Lopes, F. J. S., Montilla, E., Guedes, A. G., Hoelzemann, J. J., Fernandez, J. H., Alados-Arboledas, L., Guerrero-Rascado, J. L., 2016: DUSTER lidar: transatlantic transport of aerosol particles from the Sahara and other sources: first results from the recently installed lidar and sunphotometer in Natal/Brazil. Proc. SPIE 10006, Lidar Technologies, Techniques, and Measurements for Atmospheric Remote Sensing XII, 1000607. https://doi.org/10.1117/12.2241386.

[64] Baars, H., Ansmann, A., Althausen, D., Engelmann, R., Artaxo, P., Pauliquevis, T., Souza, R., 2011: Further evidence for significant smoke transport from Africa to Amazonia. Geophysical Research Letters: Atmospheric Science, v. 38, Issue 20. https://doi.org/10.1029/2011GL049200.

[65] Abouchami, W., Näthe, K., Kumar, A., Galer, J. G., Jochum, K. P., Williams, E., et al., 2013: Geochemical and isotopic characterization of the Bodélé Depression dust source and implications for transatlantic dust transport to the Amazon Basin. Earth and Planetary Science Letters, v. 380, 112-123. https://doi.org/10.1016/j.epsl.2013.08.028.

[66] Rizzo, L. V., Artaxo, P., Müller, T., Wiedensohler, A., Paixão, M., Cirino, G. G., et al., 2013: Long term measurements of aerosol optical properties at a primary forest site in Amazonia. Atmos. Chem. Phys., v 13, Issue 5, 2391-2413, https://doi.org/10.5194/acp-13-2391-2013.

[67] Freudenthaler, V., 2016: About the effects of polarizing optics on lidar signals and the Delta 90 calibration. Atmospheric Measurement Techniques, v. 9, Issue 9, 4181-4255. ISSN: 1867-1381. https://doi.org/10.5194/amt-9-4181-2016.

[68] Van Hai, B., Van Trung, D., Xuan Tuan, N., Duy Thang, D., Thanh Binh, N., 2012: Monitoring cirrus cloud and tropopause height over Hanoi using a compact lidar system. Communications in Physics, v. 22, Issue 4. 357-364. ISSN: 0868-3166. https://doi.org/10.15625/0868-3166/22/4/2488.

[69] Heymsfield, A. J., Krämer, M., Luebke, A., Brown, P., Cziczo, D. J., Franklin, C., et al., 2017: Cirrus clouds. Meteorological Monographs, v. 58, 2.1-2.26. https://doi.org/10.1175/AMSMONOGRAPHS-D-16-0010.1.

[70] Guedes, A. G., 2019: Characterization of the transport of Saharan dust aerosols on Natal-RN using the lidar depolarization technique [thesis] (In Portuguese). Center for Exact and Earth Sciences. Graduate Program in Climate Sciences. Federal University of Rio Grande do Norte. Available from: https://repositorio.ufrn.br/handle/123456789/27386. Accessed 6 November 2020.

[71] Boucher, O., Randall, D., Artaxo, P., Bretherton, C., Feingold, G., Forster, P., et al., 2013: Clouds and Aerosols. In Climate Change 2013: The Physical Science Basis. Contribution of Working Group I to the Fifth Assessment Report of the Intergovernmental Panel on Climate Change. Editors: Stocker, T. F., Qin, D., Plattner, G.-K., Tignor, M., Allen, S. K., Boschung, J., Nauels, A., Xia, Y., Bex, V., and Midgley, P. M., Cambridge University Press, 571-658, https://doi.org/10.1017/CBO9781107415324.016.

[72] Sassen K, Wang Z, Liu D. Cirrus clouds and deep convection in the tropics: Insights from CALIPSO and CloudSat, J. Geophys. Res., v. 114, Issue D4. In: D00H06. 2009 https://doi.org/10.1029/2009JD011916

[73] Liou, K. N., 1986: Influence of cirrus clouds on weather and climate processes: A global perspective, Mon. Weather Rev., v. 114, Issue 6, 1167 – 1199. https://doi.org/10.1175/1520-0493(1986)114<1167:IOCCOW>2.0.CO;2.

[74] Kienast-Sjögren, E., Rolf, C., Seifert, P., Krieger, U. K., Luo, B. P., Krämer, M., and Peter, T., 2016: Climatological and radiative properties of midlatitude cirrus clouds derived by automatic evaluation of lidar measurements. Atmos. Chem. Phys., v. 16, Issue 12, 7605-7621. https://doi.org/10.5194/acp-16-7605-2016.

[75] Nazaryan, H., McCormick, M. P., and Menzel, W. P., 2008: Global characterization of cirrus clouds using CALIPSO data, J. Geophys. Res., v. 113, Issue D16, D16211. https://doi.org/10.1029/2007JD009481.

[76] Pandit, A. K., Gadhavi, H. S., Venkat Ratnam, M., Raghunath, K., Rao, S. V. B., and Jayaraman, A., 2015: Long-term trend analysis and climatology of tropical cirrus clouds using 16 years of lidar data set over Southern India, Atmos. Chem. Phys., v. 15, Issue 24, 13833-13848. https://doi.org/10.5194/acp-15-13833-2015.

[77] Lakkis, G. S., Lavorato, M., and Canziani, O. P., 2009: Monitoring cirrus clouds with lidar in the Southern Hemisphere: a local study over Buenos Aires. 1. Tropopause heights, Atmos. Res., v. 92, Issue 1, 18-26. https://doi.org/10.1016/j.atmosres.2008.08.003.

[78] Antuña, J. C., Barja, B., 2006: Cirrus clouds optical properties measured with lidar at Camagüey, Cuba. Óptica Pura y Aplicada, v. 39, Issue 1, 11-16. Proceedings of the Third Workshop of Lidar Measurements in Latin America. ISSN-e 2171-8814. Available from: https://www.sedoptica.es/Menu_Volumenes/Pdfs/178.pdf. Accessed 13 October 2020.

[79] Gouveia, D. A., Barja, B., Barbosa, H. M., Seifert, P., Baars, H., Pauliquevis, T., Artaxo, P., 2017: Optical and geometrical properties of cirrus clouds in Amazonia derived from 1 year of ground-based lidar measurements. Atmospheric Chemistry and Physics, v. 17, Issue 5, 3619-3636. https://doi.org/10.5194/acp-17-3619-2017.

[80] Santos, R. S. S., 2018: Study of physical properties of cirrus clouds in

Natal using data from the lidar-Duster system and radiosondes [capstone project] (In Portuguese). Technology Center. Federal University of Rio Grande do Norte. Brazil. Available from: http://monografias.ufrn.br/handle/123456789/6908. Accessed 13 October 2020.

[81] Larroza, E. G., 2011: Characterization of cirrus clouds over São Paulo Metropolitan City (MSP) by elastic lidar [thesis] (In Portuguese). Institute of Energy and Nuclear Research. University of São Paulo. Brazil. https://doi.org/10.11606/T.85.2011.tde-19122011-153154.

[82] Barja, B. et al. 2019: Cirrus cloud observations at the southern-hemispheric midlatitude site of Punta Arenas (53°S, 71°W). International Symposium on Tropospheric Profiling (ISTP), 20 - 24 May 2019. Météo-France - Toulouse - France. http://www.meteo.fr/cic/meetings/2019/ISTP/programme_ISTP11.pdf.

[83] Barja, B., Antuña, J. C., 2011: The effect of optically thin cirrus clouds on solar radiation in Camagüey, Cuba. Atmos. Chem. Phys., v. 11, Issue 16, 8625-8634. https://doi.org/10.5194/acp-11-8625-2011.

[84] Gouveia, D. A., 2018: Radiative forcing, physical and optical properties of cirrus clouds over Amazon [thesis] (In Portuguese). Institute of Physics. University of São Paulo. Brazil. https://doi.org/10.11606/T.43.2019.tde-08022019-141530.

[85] Emde, C., Buras-Schnell, R., Kylling, A., Mayer, B., Gasteiger, J., Hamann, U., et al., 2016: The libRadtran software package for radiative transfer calculations (version 2.0.1). Geoscientific Model Development, v. 9, Issue 5, 1647-1672. https://doi.org/10.5194/gmd-9-1647-2016.

[86] Barbosa, H. M. J., Gouveia, D. A., Kalese, H., Seifert, P., Barja, B., Monteiro, M., Correia, A., Pauliquevis, T., Ansmann, A., 2018: Optical Properties and Radiative Forcing of Cirrus Clouds in the Central Amazon Forest. X Workshop on Lidar Measurements in Latin America. Medellín, Colombia. Available from: http://www.fap.if.usp.br/~hbarbosa/uploads/Site/Publications/Barbosa_wlmla2018.pdf. Accessed 3 November 2020.

[87] Larroza, E. G., Nakaema, W. M., Bourayou, R., Hoareau, C., Landulfo, E., and Keckhut, P., 2013: Towards an automatic lidar cirrus cloud retrieval for climate studies. Atmos. Meas. Tech., v. 6, Issue 11, 3197-3210. https://doi.org/10.5194/amt-6-3197-2013.

[88] Bühl, J., Seifert, P., Wandinger, U., Baars, H., Kanitz, T., Schmidt, J., et al., 2013: LACROS: the Leipzig Aerosol and Cloud Remote Observations System. Proc. SPIE v. 8890. Remote Sensing of Clouds and the Atmosphere XVIII; and Optics in Atmospheric Propagation and Adaptive Systems XVI. 889002. https://doi.org/10.1117/12.2030911.

Chapter 3

Application of Remote Sensing in Natural Sciences

Ehsan Atazadeh and Mostafa Mahdavifard

Abstract

Generally, the term biomass is used for all materials originating from photosynthesis. However, biomass can equally apply to animals. Conservation and management of biomass is very important. There are various ways and methods for biomass evaluation. One of these methods is remote sensing. Remote sensing provides information about biomass, but also about biodiversity and environmental factors estimation over a wide area. The great potential of remote sensing has received considerable attention over the last few decades in many different areas in biological sciences including nutrient status assessment, weed abundance, deforestation, glacial features in Arctic and Antarctic regions, depth sounding of coastal and ocean depths, and density mapping.

Keywords: biomass, RS, GIS, chlorophyll-a

1. Introduction

A natural resource is any kind of energy or substance that is necessary to meet the physiological, social, economic and cultural needs of humanity and to maintain all the various activities that lead to production. Natural resources such as solar energy, forests, crops, fisheries, etc. are renewable, and such as oil, coal, natural gas, etc. are not renewable (Rao). National development requires a comprehensive approach to natural resource management. Sustainable natural resource management requires long-term oversight. Periodic evaluation and monitoring of natural resources enables policymakers to be vigilant in the optimal use of resources and the development process to act promptly [1–3]. In recent years, remote sensing data has been widely used in various fields of natural resource management as it is one of the best data sources for large-scale applications [4]. In general, remote sensing is "the measurement or acquisition of information about the properties of an object or phenomenon by a recording device (sensor: satellite or aircraft) that is not studied in physical or direct contact with the phenomenon" [5]. Integrated use of remote sensing data, GPS and GIS enables consultants, natural resource managers and researchers in government agencies, conservation organizations and industry to develop management plans for a variety of natural resource management programs [6]. When natural resources are identified using remote sensing data, sampling strategies are also used to collect and evaluate field observations of the variables at the selected locations [7]. Remote sensing is a powerful tool for studying land cover, forest management, water quality parameters, etc. in remote locations. Among the studies using modern remote sensing technology to estimate the concentration of chlorophyll-A (phytoplankton pigment) in the coastal areas of Iran [8] and also in Vietnam has been used to identify and estimate the upper biodiversity of mangrove forests. In the next section, we discuss some of the applications of remote sensing in marine ecosystem management.

2. Remote sensing applications in marine ecosystems

Sea Surface temperature: Most industries use seawater as cooling water. When the water used as a coolant returns to the natural environment at a higher temperature, temperature changes reduce oxygen and affect the marine ecosystem [9]. SST information is needed in remote sensing assessment for fisheries applications [10]. Satellite SST observations help to understand regional diversity and global climate change, and allow the visualization of a wide range of ocean zones. In 1981, SST satellite infrared observations began with the launch of the AVHRR sensor on the NOAA 7 satellite, so that there are now three decades of SST satellite data [11].

Coral reef: Coral reefs are formed by living coral polyps and calcareous algae that grow at sea levels between 77 and 86 degrees Fahrenheit with normal salinity. Coral reefs act as feeding, reproductive, breeding, and aquatic habitats for many oceanic organisms, so the density and distribution of corals alter the populations of fish and other organisms [12]. The advantages of remote sensing systems in this area are the ability to survey the area on a large scale, continuous monitoring and the ability to check the system ecosystem in remote areas without direct contact

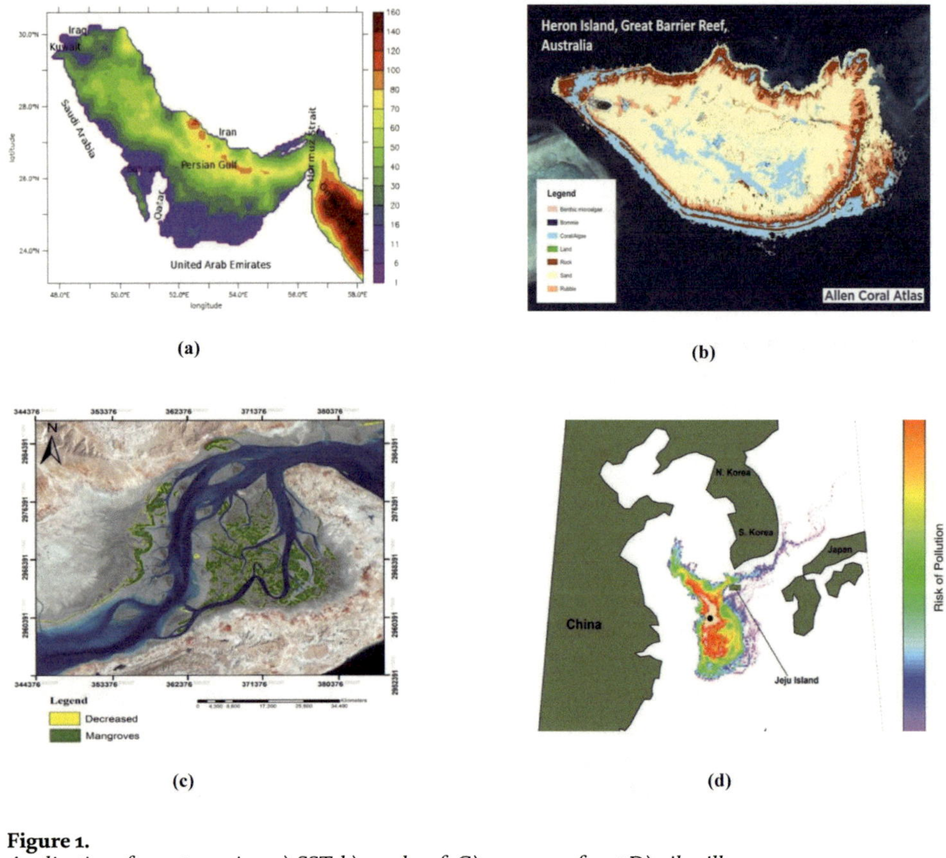

Figure 1.
Application of remote sensing: a) SST, b) coral reef, C) mangrove forest D) oil spill.

with it. Among these, optical Remote sensing systems including multispectral and hyperspectral sensors have been most used in this field [13, 14].

Mangrove forest: Mangrove forests are tropical and subtropical ecosystems that grow on the margins of two different environments, sea and land [15–17]. These forests play an essential role in ecology and are considered as carbon reserves [18–20]. Due to their location in the tropics, they are prone to hiding under clouds, so a reliable monitoring tool is essential to measure deforestation. Radar remote sensing has been shown to be useful in distinguishing mangrove cover from other ground cover due to its long wavelengths that can penetrate the cloud [21].

Ocean Color: The color of the ocean is a unique property for water. Mapping and understanding ocean color changes can help monitor water quality and identify natural and human contaminants (oil slicks and algae blooms) that are dangerous to aquaculture and even to humans [8]. Remote sensing optical data can detect targets such as suspended sediments, algal blooms, chlorophyll-A, and oil slicks at various scales [21]. In the next section, which is research-oriented, we want to examine the application of remote sensing in estimating the concentration of chlorophyll- a phytoplankton in coastal waters (**Figure 1**).

3. Materials and methods

3.1 Case study

The current research, which is the result of the authors' efforts, has been conducted on the Tiab estuary. Tiab estuary in Hormozgan province - Iran is one of the estuaries of Sirik and Kolahy basin which is located in latitudes of 27 degrees and 8 minutes and longitude of 56 degrees and 44 minutes (**Figure 1**). The climatic conditions of this region are tropical to subtropical and its climate is hot and humid. The temperature in this region reaches 45 degrees Celsius in summer. The average annual rainfall in this region is about 100 to 300 mm between November and April. Most of the vegetation of this area includes pure communities of mangroves (Avicenina marina) which is located at the mouth of the estuary and the edges of tidal waterways (**Figure 2**) [22].

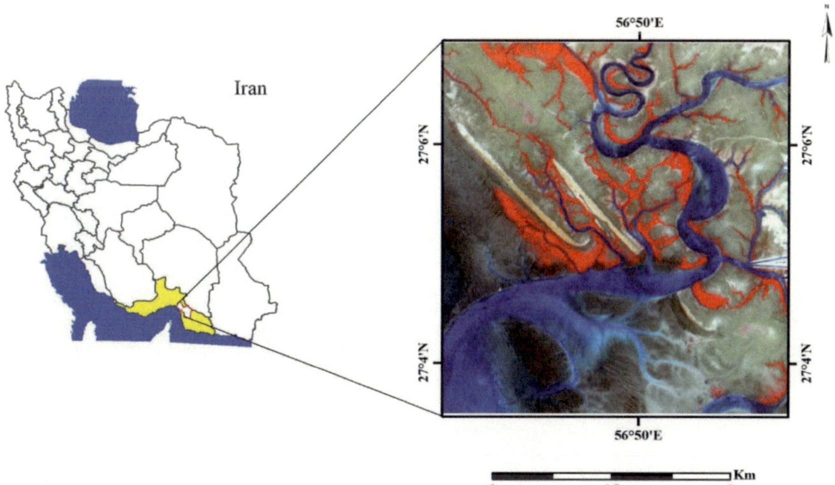

Figure 2.
Location of the study area.

3.2 Data collection

3.2.1 In-situ data

To collect field data from the study area for calibration with remote sensing data, field measurements were performed on April 7, 2019, at the same time as the Landsat-8 satellite passed the study area at 6:37 UTC. On the day of sampling, the sea level was at the highest water level and without wave turbulence. For this purpose, immediately after collecting the water sample using special bottles, the location of each station was determined using a GPS device. A total of 10 stations were located and sampled. The sampling started from the area between the entrance of the estuary and ended at the exit of the estuary leading to the high seas and lasted about 1 hour. **Figure 3** shows the field operation. In the laboratory, water samples were transferred to tubes containing 10 ml of ethanol and then centrifuged at 2500 rpm for 5 minutes and decomposed in the absorption range of 664 and 665 nm using a spectrophotometer. Finally, the field chlorophyll of each station was calculated using Eq. 1.

$$\mathrm{Chla}_{\left(mg/\mathrm{m}^3\right)} = \frac{26.7 \times \left(664_{\mathrm{b}} - 665_{\mathrm{a}}\right) \times \mathrm{V}_{ext}}{\mathrm{V}_{sam} \times \mathrm{L}} \tag{1}$$

Where 665a and 664b, respectively, before and after acidification of the samples in the spectral range, Vext amount of extraction per liter unit, Vsam amount of sample per cubic meter, L light path of the sample tube per unit centimeter and Chla chlorophyll concentration per mg unit Cubic meters.

3.2.2 Remote sensing data

In this study, Landsat-8 satellite images were used to estimate chlorophyll-A concentration. Landsat-8 is one of the newest multispectral satellites launched on February 11, 2013 to monitor natural resources. This satellite has a spatial resolution of 30 meters, 11 Multispectral bands (in the range: visible, reflective infrared and thermal infrared) and its review period from Earth is 16 days. **Table 1** shows the details of Landsat-8 time series data for estimating chlorophyll-A concentration are presented.

Figure 3.
Field operations in the study area.

ID-Landsat-8	Flight time	Date
LC08_L1TP_159041_20190407_20190407_01_RT.tar	6:39 (UTC)	2019/4/7
LC08_L1TP_159041_20190728_20190801_01_T1.tar		2019/7/28
LC08_L1TP_159041_20191117_20191202_01_T1.tar		2019/11/17

Table 1.
Specifications of time series data used.

3.3 Modeling of chlorophyll-A concentration

Bio-optical algorithms OC2 and OC3 were used to extract chlorophyll-A concentration from Landsat-8 data. This bio-optical model is based on nonlinear relationships between ocean reflection reflectance and field data that actually link optical measurements of reflection to qualitative parameters such as chlorophyll concentration and range-of-coastal, blue, and green weather to estimate chlorophyll concentration. Uses the following equations to calculate.

$$C = 10^{(a0+a1*R+a2*R2+a3*R3)}$$

$$R = \log(Rrs\ Blue\ /\ Rrs\ Green) \tag{2}$$

$$C = 10^{(a0+a1*R+a2*R2+a3*R3)}$$

$$R = \log(Rrs\,490\ /\ Rrs\,555) \tag{3}$$

Where **a** are coefficients that are estimated and modeled through the relationship between field data and remote sensing. **Rrs** is a measure of distance at blue and green wavelengths, **R** is the band ratio and **C** is the concentration of chlorophyll-A in milligrams per cubic meter.

4. Results and analysis

4.1 Field measurement

The results of field sample analysis showed that the lowest concentration of chlorophyll-A was in station number 9 with 0.11 mg/m^3 and the highest level was in station number 6 with 6 mg/m^3 (due to severe algae) and in the rest of the stations. Chlorophyll content is below 1 mg/m^3. The diagram below shows the trend of chlorophyll-A changes in the stations (**Figure 4**).

4.2 Chlorophyll-A validation received from satellites with field data

The results of chlorophyll-A evaluation show that the use of OC2 bio-optical algorithm to estimate chlorophyll-A concentration using Landsat-8 data has acceptable results and has a high correlation with field data. **Table 2** shows the result of statistical parameters shows the correlation between field data and chlorophyll derived from Landsat-8 data.

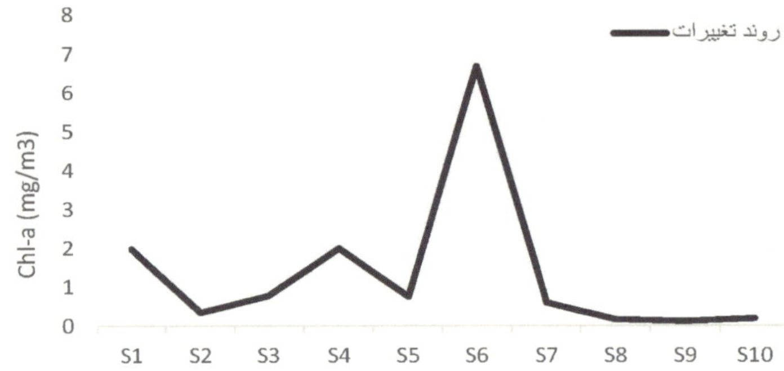

Figure 4.
The trend of field chlorophyll changes.

Algorithm	R^2	RMSE	a_x
OC2	0.91	0.13	−0.20094
			−1.3018
			0.52631
			1.4235
OC3	0.88	0.16	−0.22314
			−1.3116
			0.70002
			1.4329
			−0.16545

Table 2.
Results of RMSE, R2 and Landsat-8 modeled coefficients with field data.

4.3 Seasonal estimation of chlorophyll-A concentration with Landsat-8

After obtaining the appropriate results from the OC2 algorithm in Landsat-8 Remote Sensing data using Eq. (2) and the modeled coefficients presented in the table above, a chlorophyll-A concentration map was prepared in the seasons of 2019. The chlorophyll-A concentration estimation map is shown in **Figure 5**.

$$C = 10^{(a0+a1*R+a2*R2+a3*R3)}$$

$$R = \log\left(Rrs\,490 \,/\, Rrs\,555 \right) \tag{4}$$

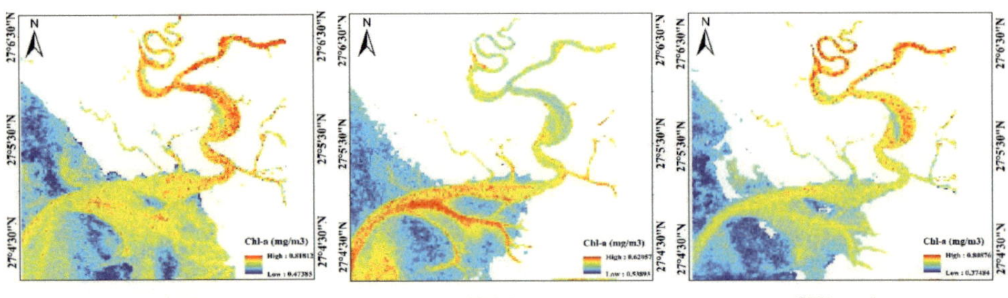

Figure 5.
Time series map of chlorophyll-A concentration in Tiab estuary prepared from Landsat-8 images [22].

The coefficients a0, a1, a2, a3, a4 were determined for Landsat OC2 algorithm −0.20094, −1.3018, 0.52631 and 1.4235, Rrs is the reflection of distance measurement at blue and green wavelengths, R is the band ratio and C, chlorophyll-A concentration to The unit is milligrams per cubic meter.

5. Discussion and conclusion

In this study, the potential of remote sensing data for modeling and estimating the time series of chlorophyll-A concentration with field data for the coastal region of Khor Tiab was evaluated.Coastal Aerosol, Blue and green band-based bio-optical models (OC2 and OC3) were applied to Landsat-8 satellite multispectral data to estimate chlorophyll-A concentrations in the coastal area. The results of this study show that OC2 and OC3 bio-optical models have a high correlation with field data sampled from the coastal waters of Tiab estuary so that the statistical parameter values R2 and RMSE in the OC2 model with field data were equal to 0.91 and 0.13, respectively. While the results of statistical parameter R2 and RMSE in the OC3 model were evaluated with field data with a slight difference of 0.88 and 0.16, respectively. Due to the good performance of the OC2 algorithm compared to the OC3 algorithm with field data, this algorithm was selected and used to estimate the time series of chlorophyll-A concentration in the Tiab estuary. As the time series map shows, the concentration of chlorophyll-A in April and November compared to July has a relatively high concentration of chlorophyll that follows the natural conditions, which is due to the fact that in April and November the sea temperature is low and cold. This has increased the concentration of chlorophyll-A in the region, while the concentration of chlorophyll-A has decreased significantly in July because the sea temperature is high this month and is completely correlated with natural conditions. In fact, there is an inverse relationship between chlorophyll-A concentration and sea surface temperature, which fully illustrates the Remote Sensing data used in this study. The end result of this study indicates that remote sensing data is a much better alternative to field operations due to its cost and time consuming nature. Finally, the participants of this study are suggested to use more field data to calibrate the Remote Sensing data in order to estimate the chlorophyll-A concentration in order to obtain better results.

Author details

Ehsan Atazadeh* and Mostafa Mahdavifard
University of Tabriz Tabriz, Iran

*Address all correspondence to: islam.atazadeh@gmail.com

References

[1] Gaur M, Pandey CB, Goyal RK. Remote Sensing for Natural Resources Management & Monitoring. Scientific Publishers; 2016

[2] http://www.felixstowedocker.com/2018/01/sanchis-sinking-raises-fears-of.html

[3] https://www.rsrc.org.au/news/2018/10/31/unprecedented-global-image-of-coral-reefs-from-allen-coral-atlas-released

[4] Kumar N, Yamaç SS, Velmurugan A. Applications of remote sensing and GIS in natural resource management. Andaman Science Association. 2015;**20**(1):1-6

[5] Colwell RN, editor. Manual of Remote Sensing, 2nd. Ed. Falls Church: ASP&RS; 1983

[6] Philipson P, Lindell T. Can coral reefs be monitored from space? Ambio. 2003;**32**:586-593

[7] Wang G, Weng Q, editors. Remote Sensing of Natural Resources. CRC Press; 2013

[8] Mahdavifard M, Valizadeh Kamran K, Atazadeh E. Estimation of chlorophyll-a concentration using ground data and Sentinel-2 and Landsat-8 satellite images processing (case study: Tiab estuary). Journal of RS and GIS for natural. Resources. 2020;**11**(1):72-83

[9] Cahyono, A. B., Saptarini, D. I. A. N., Pribadi, C. B., & Armono, H. D. (2017). Estimation of Sea Surface Temperature (SST) Using Split Window Methods for Monitoring Industrial Activity in Coastal Area. In Applied Mechanics and Materials (Vol. 862, pp. 90-95). Trans Tech Publications Ltd.

[10] Nontji, A. (1993). Laut Nusantara Penerbit Djambatan.

[11] Martin S. An Introduction to Ocean Remote Sensing. Cambridge University Press; 2014

[12] Nybakken, J. W., & Bertness, M. D. (2001). Marine Biology: An Ecological Approach 5th Edition.

[13] Andréfouët S, Kramer P. Multi-site evaluation of IKONOS data for classification of tropical coral reef environments. Remote Sensing of Environment. 2003;**88**:128-143

[14] Mumby, P.; Green, E.; Edwards, A.; Clark, C., 1997. Coral reef habitat mapping: How much detail can remote sensing provide? Marine Biology, 130: 193202.

[15] Rao, P. N. Applications of Remote Sensing and Geographical Indication System in Land Resources Management.

[16] Rezaei-Latifi A, Hosseinibalam F. An estimate of the surface heat fluxes transfer of the Persian Gulf with the overlying atmosphere. Journal of Radiation Research and Applied Sciences. 2015;*8*(3):354-361

[17] Saleh M. Assessment of mangrove vegetation on Abu Minqar Island of the red sea. Arid Envirmental. 2007;**68**:331-336

[18] Kathiresan K, Bingham BL. Biology of mangroves and mangroves ecosystems. Advances in Marine Biology. 2001;**40**:81-251

[19] Toosi NB, Soffianian AR, Fakheran S, Pourmanafi S, Ginzler C, Waser LT. Comparing different classification algorithms for monitoring mangrove cover changes in southern Iran. Global Ecology and Conservation. 2019;**19**:e00662

[20] Valiela I, Bowen JL, York JK. Mangrove forests: One of the world's

threatened major tropical environments. Bioscience. 2001;**51**:807-815

[21] CCRS, C. f. RS, (2014). Tutorial: Fundamentals of Remote Sensing.

[22] Mahdavifard M. Estimate of Chlorophyll-a Concentration Using In-Situ and Satellite Image Processing Approach: Estuary of Tiab. University of Tabriz: Master's thesis; 2020

Chapter 4

Assessment of Ecological Disturbance Caused by Flood and Fire in Assam Forests, India, Using MODIS Time Series Data of 2001-2011

Dibyendu Dutta, Akanksha Balha, Prabir Kumar Das, Pragyan Jain, Libeesh Lukose and Wasim Akram

Abstract

The forest area of Assam State is known for its rich biodiversity. In the present study, the disturbance regime within the Assam forest area caused by periodic flood and forest fire, was assessed using the Moderate Resolution Imaging Spectroradiometer (MODIS) time-series (2001–2011) data. The MODIS Global Disturbance Index (MGDI) images were generated using MODIS derived Enhanced Vegetation Index (EVI) and Land Surface Temperature (LST) images. The temporal intensity of flood and forest fire in sixteen representative forests was analyzed to develop the MGDI based thresholds for detecting the disturbed area. The threshold for the non-instantaneous disturbance, i.e. flood, was found to be 107% whereas it was 111% for instantaneous disturbance, i.e. forest fire. The thresholds were applied on the MGDI images to delineate disturbed caused by flood and fire, separately for each year. The time-series disturbance areas were integrated over the years (2001–2011) to generate the classified disturbance prone maps.

Keywords: ecological disturbance, MGDI, flood, forest fire, Assam forest, MODIS, EVI, LST

1. Introduction

The north-eastern state of Assam is known for its rich biodiversity and considered as biological hotspot with many rare and endemic plant and animal species. Out of total 78,438 sq. km geographical area of the state, the forest area covers around 24.58% area. The moderate dense forests areas which are mainly extended through districts of Karbi Anglong, NC Hills, Cachar, Karimganj, Hailakandi, northern part of Kokrajhar, Bongaigaon and southern part of Kamrup, Tinsukia, are vulnerable to frequent flood incidents. The Brahmaputra River and its tributaries, flowing from north-east to south-west, are the mainly responsible for the periodic floods in Assam State. Along with flood events, the incidence of forest fires in the deciduous forests during summer season, i.e. March to April, causing a wide spread disturbance in the forest eco-system. Assam State has 5 National Parks and 16 Wildlife Sanctuaries under protected area (PA) network and constituting 4.98% of the geographical area. The protected areas can act as benchmark for differentiating the ecological disturbance from the natural fluctuation [1]. Hence, regular monitoring of PA's is crucial for detecting the rapid changes in functional attributes as well as to identify areas that need to adapt or modify to meet the challenges posed by global warming [2].

The understanding of the global carbon cycle is being affected due to the existing spatio-temporal variability of eco-system disturbance and resultant emissions from loss of terrestrial biomass [3–8]. Hence, the regular monitoring and assessment of the ecological disturbance is essential for understanding the cause and effect of the disturbances and subsequent effective management of the forest ecosystems. With the advent of multispectral and thermal remote sensing technology, the earth observation satellites data became more effective for monitoring biodiversity. The altered ecosystem structure and functions due to sustained disturbance in Woody ecosystem can be captured by remote sensing for mapping the extent and location of the disturbance [9, 10]. The effectiveness of the management practices or impact of global environment changes in the forest areas, especially Protected Area (PA), can be successfully carried out using satellite remote sensing [11, 12]. The technology can also provide valuable information on the alteration of landuse, productivity or phenology [13].

The MODIS global disturbance index (MGDI) was first proposed and used by [14] for assessing the disturbance in woody ecosystems of North America. The MGDI was conceptualized based on the fact that the surface temperature decreases with the increase in vegetation density through the latent heat transfer [15, 16]. The index was further utilized for assessing the impact of cyclones on the ecological disturbance of mangrove forest of Sundarbans. For further details about the concept and computation of MGDI, one can refer [14] or [16].

In the present study, MODIS global disturbance index (MGDI) was used to assess the ecological disturbance caused by two different kinds of natural hazards, viz. river flood and forest fire, in the perennial forest ecosystem of Assam State during 2001 to 2011. The flood and forest fire induced disturbed areas were identified based on the MGDI based thresholds and the spatio-temporal dynamics of the disturbance over the Assam forest area was studied. Finally, a classified geo-spatial product of disturbed prone forests was generated based on estimated disturbed area during the entire study period.

2. Study area

Assam State, situated at the north-east of India and foothills of the eastern Himalayas, covers an area of 78,438 sq. km and lies in the middle reach of the Brahmaputra River and Barak (http://www.asbb.gov.in/geophysical.html). The state is bounded by 88.25°E to 96.00°E longitude and 24.50°N to 28.00°N latitude (**Figure 1**). Mean annual temperature varies from 6–38°C and average annual rainfall is 3000 mm on the Brahmaputra River valley and the surrounding region. In Assam State around 51 types of forests and sub-forests can be found. Physiographically the state can be classified into 3 groups, viz. vast alluvial plains of the Brahmaputra River valley in the north, the central Assam hills and the hilly and alluvial terrain in the south.

In the present study, the ecological disturbance regime of the Assam forest area, India has been analyzed. Based on the intensity of flood and fire incidence, total 16 forests have been selected for MGDI based thresholds development and

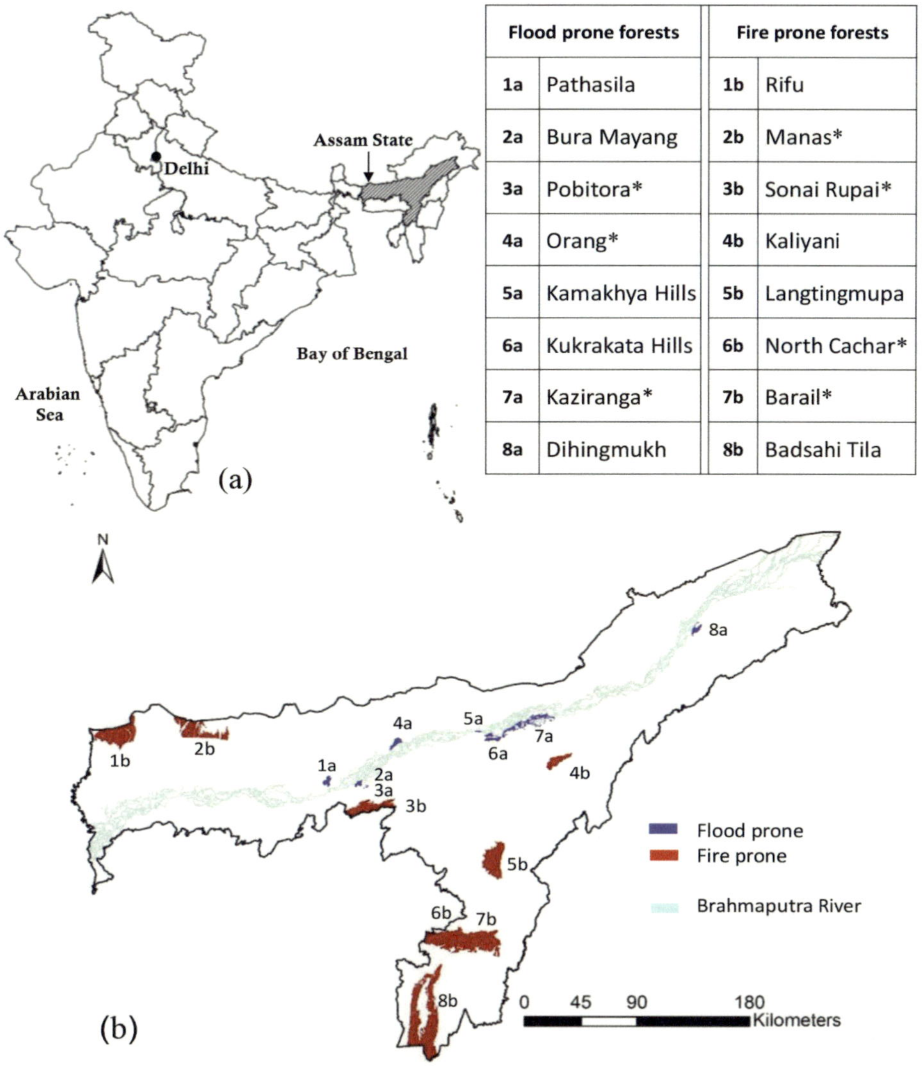

Flood prone forests		Fire prone forests	
1a	Pathasila	**1b**	Rifu
2a	Bura Mayang	**2b**	Manas*
3a	Pobitora*	**3b**	Sonai Rupai*
4a	Orang*	**4b**	Kaliyani
5a	Kamakhya Hills	**5b**	Langtingmupa
6a	Kukrakata Hills	**6b**	North Cachar*
7a	Kaziranga*	**7b**	Barail*
8a	Dihingmukh	**8b**	Badsahi Tila

Figure 1.
(a) Location of the study area in India (b) flood and fire prone forests.

discrimination of the disturbed areas. The details of the selected forests, i.e. eight for each of flood and forest fire, were shown in **Figure 1**. The flood prone forests were mainly situated along the Brahmaputra River, whereas the fire prone forests were distributed throughout the Assam State.

3. Data used

3.1 Satellite data

The 16-day composite Enhanced Vegetation Index (EVI) data products (MOD13Q1) and 8-day composite Land Surface Temperature (LST) data products (MOD11A2) for the period of 2001 to 2011 were downloaded from the MODIS web-site (www.e4ftl01.cr.usgs.gov/MOLT/). The datasets were re-projected to Geographic (Lat/Long) projection and respective scale factors were applied on the datasets [16]. The 8-day composite LST data was converted to 16-day composite using simple average, as the EVI is available at 16-day interval.

3.2 Forest boundary

The forest boundaries of Assam State were generated by digitizing the Survey of India Topomaps at 1:50,000 scale. According to the World Database on Protected Areas (www.protectedplanet.net) Assam State constitutes total 14 Protected Areas (PA), out of which 7 viz. Pabitora, Orang, Kaziranga, Manas, Sonai-Rupai, North Cachar and Barail, were analyzed in the present study. Theoretically no resource exploitation is allowed in PA's of categories I and II (IUCN, 1994). For further analysis, boundary pixels were excluded to avoid the contamination error.

3.3 Flood inundated area

The date wise flood maps were downloaded from National Remote Sensing Centre (NRSC) web site (www.nrsc.gov.in) prepared by Disaster Management Support Service Group of NRSC/ISRO, Hyderabad. The extent of flood was extracted from each geo-referenced image and stacked at different time scale (annual and multi-year). Flood intensity maps were generated based on the number of flood occurrence in a pixel within a year (**Figure 2**).

3.4 Forest fire data

Date wise forest fire information for Assam State was collected from MODIS site (https://earthdata.nasa.gov/data/near-real-time-data/firms/active-fire-data) for different locations as point information. A grid of 1 × 1 km dimension was created for the entire forest area and total number of forest fire incidents within each grid was recorded at different temporal interval (annual and multi-year) to generate fire intensity grid and subsequent use (**Figure 3**). Major districts where forest fire was reported during the study period include Hailakandi, Cachar, Karbi Anglong, Kamrup, Kokrajhar and NC Hills. The forest fire frequency for each of the selected forests during the study period was given in **Table 1**.

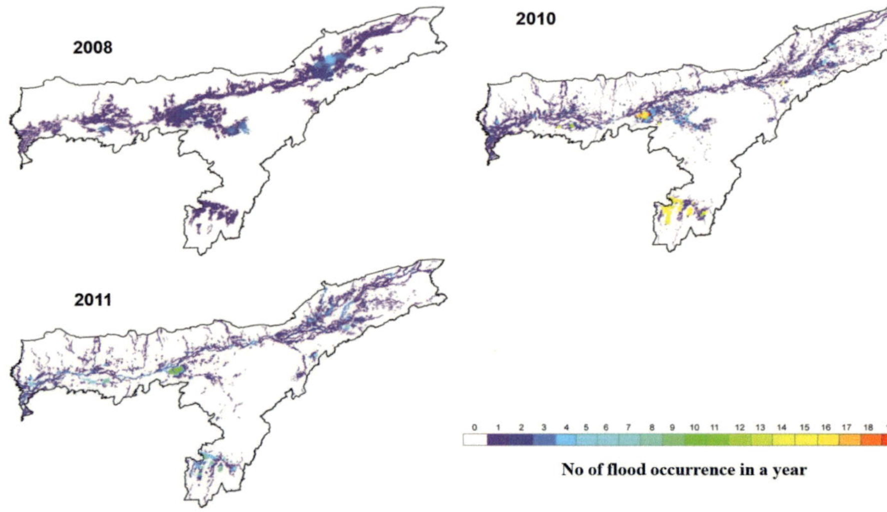

Figure 2.
Flood intensity map of Assam State (www.nrsc.gov.in).

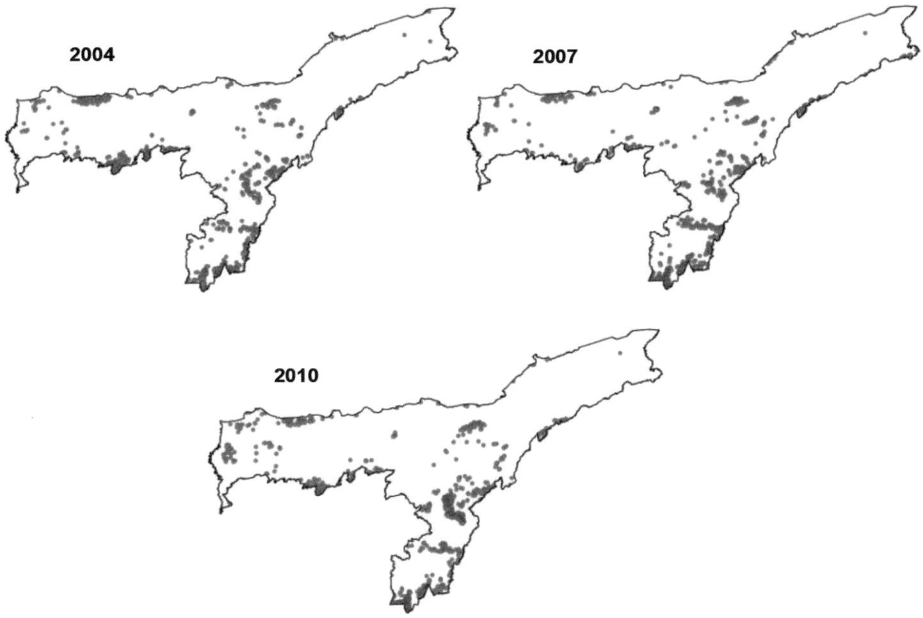

Figure 3.
Forest fire intensity map.

Year	Rifu	Manas	Sonai Rupai	Kaliyani	Langting Mupa	North Cachar	Barail	Badsahitila
2001	15	1	10	0	1	1	2	18
2002	2	23	1	1	2	0	1	16
2003	4	18	18	4	20	19	40	219
2004	12	17	55	5	44	20	26	96
2005	12	6	18	6	64	11	25	86
2006	35	13	32	6	41	31	58	158
2007	6	6	19	12	43	17	45	252
2008	4	4	41	16	25	7	27	108
2009	52	24	54	18	71	33	84	157
2010	11	17	21	4	165	11	41	212
2011	36	23	29	2	64	16	14	91

Table 1.
Forest fire events for selected Assam State forests (2001–2011).

4. Methodology

The MODIS global disturbance index (MGDI) was developed based on the concept that any perceptible disturbance of ecology will result in a significant alteration in vegetation and a concomitant change in the land surface temperature [14, 16].

In the present study, the flood and forest fire were selected as the causative factors that create ecological disturbance to address both the non-instantaneous and instantaneous disturbance, respectively. In case of instantaneous disturbance like forest fire, the disturbance is manifested immediately after the event, resulting in immediate increase in LST with decreased vegetation cover. On the contrary, non-instantaneous disturbance like flood will not trigger immediate change in either LST or EVI due to availability of abundant moisture for evaporation to offset the loss of transpiration. Whereas, in the following year the effect of flood damage was evident due to vegetation mortality and severe structural damage, which will eventually lead to increase in annual maximum LST due to the reduction in transpiration [14].

The instantaneous ($MGDI_{inst}$) and non-instantaneous ($MGDI_{non\text{-}inst}$) MGDI were computed using the following equations:

$$MGDI_{inst} = \frac{\left(LST_{max} / EVI_{max-post}\right)_{current\ year(y)}}{\left(LST_{max} / EVI_{max-post}\right)_{multi-year\ mean(y-1)}} \quad (1)$$

$$MGDI_{non-inst} = \frac{\left(LST_{max} / EVI_{max}\right)_{current\ year(y)}}{\left(LST_{max} / EVI_{max}\right)_{multi-year\ mean(y-1)}} \quad (2)$$

Where, $MGDI_{inst}$ and $MGDI_{non\text{-}inst}$ are the instantaneous and non-instantaneous MGDI value, respectively. LST_{max} and EVI_{max} are the annual maximum 16-day composite LST (°C) and EVI, respectively. $EVI_{max\text{-post}}$ is the maximum 16-day composite EVI following the LSTmax, current year (y) is the year being evaluated for disturbance and multi-year mean (y − 1) is the mean of the ratios excluding the current year [14, 16].

A two-step methodology, as explained by Dutta et al. [16], was adopted for to discriminate the disturbed forest areas caused due to flood and forest fire. In the 1st step, the % change in MGDI values were calculated based on the time-series data for each pixel. Whereas in the second step the MGDI based thresholds were estimated for flood and forest fire, separately. The Percentage change in MGDI for both the instantaneous and non-instantaneous disturbance were calculated using the following equation:

$$\%change\ in\ MGDI_{current\ year(y)} = \frac{MGDI_{current\ year(y)}}{Multi-year\ mean\ MGDI_{(y-1)}} \times 100 \quad (3)$$

As discussed earlier, the forests prone to flood and fire were selected based on the temporal occurrence of the natural disturbance. Total 16 representative forests frequently affected by flood and forest fire were extensively analyzed to develop the thresholds for flood and forest fire separately based upon the % change of MGDI over multi-year mean. The spatio-temporal variation of the % change in $MGDI_{non\text{-}inst}$ over the Assam forest area was shown in **Figure 4** for some selected years.

The year-wise % change in MGDI was generated for all the representative forests wherein only the flood affected pixels were considered. Similarly, year-wise % change in MGDI was generated for the pixels undergone forest fire. The temporal profile of the percent change in MGDI of each forest was compared with the area weighted flood and fire intensity, to confirm the effect of natural disturbances on the MGDI. The multi-year mean value plus one standard deviation of the % change in MGDI was considered to be the threshold, and the value was used for

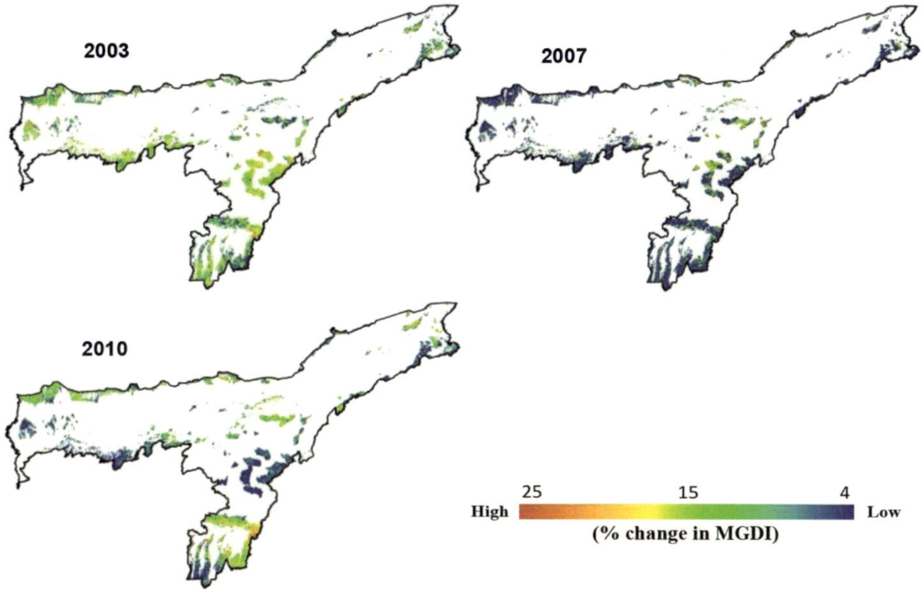

Figure 4.
Percent change in $MGDI_{non\text{-}inst}$ over the Assam forest area.

discrimination of the disturbed pixels [16]. Due to the slow and gradual impact of flood, unlike forest fire, a lower disturbance threshold was estimated in case of flood. The % change of MGDI greater than 7% and 11% of the temporal mean was fixed for discriminating the non-instantaneous (flood) and instantaneous (forest fire) disturbed pixels.

The selected thresholds were applied on the % change MGDI images (both non-instantaneous and instantaneous, separately) for identifying the year-wise disturbed forest areas. The year-wise % disturbed area was estimated for each forest and the temporal profiles were used to analyze the disturbance intensity at spatio-temporal scale. Upon integration of the year-wise disturbed area, disturbance prone maps were generated for both instantaneous and non-instantaneous events.

5. Results and discussion

5.1 Disturbed area caused by the flood

The summary statistics of the flood affected forests is shown in **Table 2**. According to the % disturbed area it was found that the year 2003, 2008, 2009 and 2010 were the most disturbed years caused by river flood in comparison to rest of the years. These findings are in corroboration with the rainfall data collected from Indian Meteorological Department (www.imd.gov.in). During 2001, 2002, 2004, 2005 and 2006, the disturbed area for all the selected forests were less than 10% of the total area, whereas in 2007 and 2011, only few of the selected forests like Pobitora, Orang and Kukrakata Hills could cross the disturbed area limit of 10%. During 2003 the impact of flood was more than 2010 in the forests namely Phathasil, BuraMayang and Orang. On the contrary, in case of Pobitora, Kamakhya Hills, Kukrakata, Kaziranga and Dihing mukh the devastating effect of the 2010 flood was found to supersede the effect of 2003 flood.

The disturbed area maps of the Assam forests showed (**Figure 5**) that the area under disturbance was much higher in the year of 2003, 2008 and 2010 in comparison to others, whereas it was minimum in the year of 2006. The flood intensity maps (**Figure 2**) showed that the extent of flooded area is mainly confined around the Brahmaputra and Barak River valley, but it was interesting to note that a subset of the flood pixels were marked as disturbed pixels (**Figure 5**). Hence, the temporal frequency of flood along with extent was a decisive factor whether a pixel was disturbed or not. For example, in BuraMayang around 80% of the total forest area

Year	Phathasil	Bura Mayang	Pobitora	Orang	Kamakhya Hills	Kukrakata Hills	Kaziranga	Dihing mukh
2001	8.4	1.2	4.7	8.4	6.7	2.0	1.4	1.4
2002	7.4	9.1	8.4	4.3	7.9	2.0	3.4	4.3
2003	39.1	31.2	17.1	10.4	29.6	9.3	4.0	6.1
2004	7.2	5.4	4.0	3.0	7.5	4.1	0.7	1.4
2005	0.5	1.1	0.0	2.6	0.0	0.0	0.3	3.3
2006	0.4	1.5	0.9	3.6	2.5	0.0	0.1	2.9
2007	4.8	8.5	14.0	4.1	8.8	1.1	2.8	5.8
2008	18.9	9.0	2.5	10.0	12.9	17.8	0.9	7.4
2009	28.0	10.2	10.2	12.4	2.5	6.1	2.6	9.4
2010	12.2	24.0	30.1	8.1	34.6	22.8	14.2	20.6
2011	1.9	1.2	0.6	11.8	2.5	11.1	2.0	2.1

Table 2.
Percent disturbed area caused by flood in selected Assam State forests.

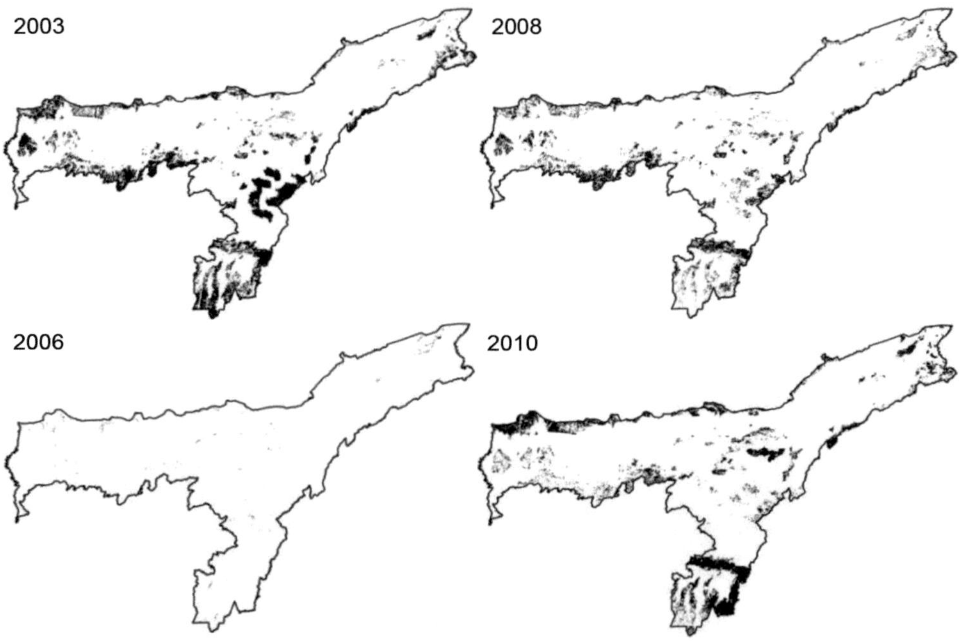

Figure 5.
The disturbed area maps of Assam forests shows higher disturbances in the 2003, 2008 and 2010 and a minimum in 2006.

was under flood during 2008 with frequency value of one, whereas during 2010 around 40% forest area was inundated by flood with frequency more than 10. Consequently, the disturbance was recorded in 9% of the total forest area in 2008, whereas it was around 24% during 2010. Similarly, in case of Pobitora the whole forest was under flood during both 2008 and 2010, but due to the difference in flood frequency the disturbed area were 2.5 and 30.1% of the total area, respectively.

The temporal dynamics of disturbed area, caused by river flood, for two selected forests, namely Kamakhya and Kukrakata Hills was shown in **Figure 6**. In case of Kamakhya Hills the year 2010 was found to the most disturbed year, followed by 2003 and 2008. Hardly any disturbance was noticed during 2005, 2006, 2009 and 2011. Similarly, in case of Kukrakata Hills also three major flood events were observed in 2010, 2008 and 2003. Though the major flood years were common in both the cases, their magnitude varies.

5.2 Disturbed area caused by forest fire

The distribution of disturbed area caused by forest fire is shown in **Table 3**. Unlike flood, the spatial extent of % disturbed area caused by forest fire was much lower as the fire is a localized phenomenon. A maximum value of 3.5% of the total forest area was disturbed due to the fire during 2010 in Barail forest. More than 2% of the forest area was affected in Sonai Rupai, Langting Mupa, Barail and Badsahitila forests during 2003. Whereas in 2010 more than 1% of the total forest area of Sonai Rupai, North Cachar and Barail was affected.

It was noteworthy that unlike flood, forest fire intensity was not in direct corroboration with disturbed area statistics (**Tables 1** and **3**). As discussed earlier, the point locations of forest fire were converted to fire intensity information using 1 km grid, to make it spatially contiguous, as the extent of forest fire information was not available. Hence, the discrepancy between the disturbed area statistics and fire frequency may be attributed to lack of spatial representation of fire extent. In addition frequent fire incidences might have hindered the process of re-generation and vegetative growth of the forest causing insignificant changes in the MGDI values in post incidence dates. For example, in case of Rifu forest the fire frequency was 52 and 36 during 2009 and 2011, respectively, but the disturbed area was 0.34 and 0.45% of the total forest area. On the other hand, 0.84% of the forest area was found to be disturbed during 2010 with only 11 fire incidences. Similar kind of observation was found in case of Badsahitila also, where the fire frequency was

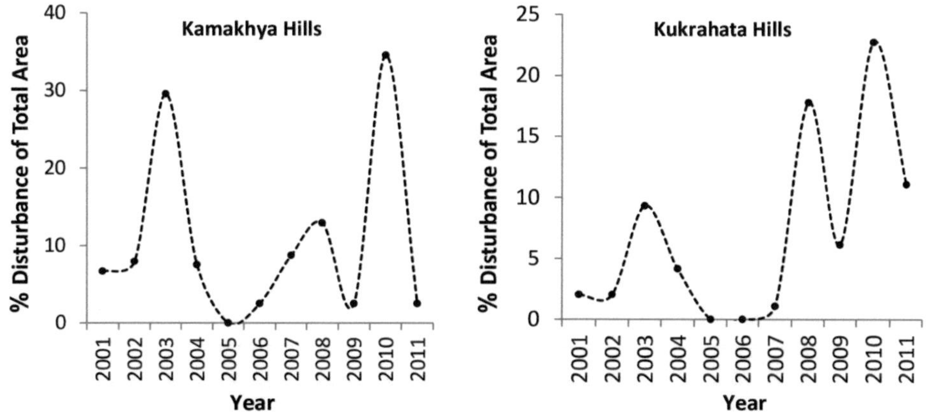

Figure 6.
Temporal dynamics of disturbed area caused by river flood for two selected forests of Assam state.

Year	Rifu	Manas	Sonai Rupai	Kaliyani	Langting Mupa	North Cachar	Barail	Badsahitila
2001	0.18	0.00	0.05	0.00	0.01	0.01	0.00	0.14
2002	0.00	0.00	0.03	0.26	0.18	0.00	0.00	0.12
2003	0.28	0.26	2.11	0.53	1.80	0.98	1.88	3.73
2004	0.24	0.23	0.07	0.13	0.98	0.08	0.07	0.73
2005	0.01	0.01	0.67	0.13	0.15	0.16	0.60	0.34
2006	0.01	0.01	0.00	0.00	0.20	0.11	0.16	0.00
2007	0.00	0.00	0.23	0.39	0.29	0.25	0.48	0.28
2008	0.07	0.08	1.78	0.34	0.26	0.31	0.58	0.05
2009	0.34	0.46	0.46	0.00	0.28	0.31	0.23	1.39
2010	0.84	0.44	1.55	0.05	0.00	1.34	3.44	0.69
2011	0.45	0.17	0.23	0.06	0.06	0.12	0.44	0.11

Table 3.
Year-wise percentage disturbed forest area caused by forest fire for selected forests of Assam State.

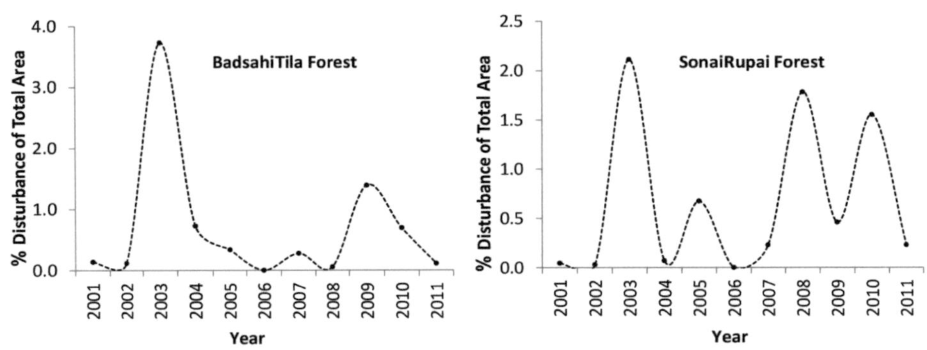

Figure 7.
Temporal dynamics of disturbed area caused by the forest fire for selected forests of Assam state.

more than 150 times during 2003, 2006, 2007, 2008, 2009 and 2010, but the % disturbed area were 3.73, 0.21, 0.28, 0.05, 1.39 and 0.69, respectively.

The temporal dynamics of disturbed area induced by the forest fire for two selected forests, namely Badsahitila and Sonai Rupai, were shown in **Figure 7**. In case of Badsahitila, the major disturbance due to forest fire was found during 2003, followed by the year 2009, whereas during 2001, 2002, 2005, 2006, 2007, 2008 and 2011 very less area was noted to be disturbed due to the fire incidences. On the other hand, in case of Sonai Rupai three major disturbances were noted during 2003, 2008 and 2010, with two other intermediate disturbances during 2005 and 2009. The major fire incidents were reported from moist deciduous forest and grass lands.

5.3 Mapped forest areas prone to disturbance

The spatial distribution of both the non-instantaneous and instantaneous forest disturbances maps have been generated and presented in **Figures 8** and **9** respectively. In non-instantaneous disturbance the effect is not triggered immediately in terms of changes in LST and/or EVI. In contrast post-event effect is immediately exhibited due to changes in LST and EVI, for example, in case of forest fire both the LST increases and the EVI changes drastically. Based upon the percent

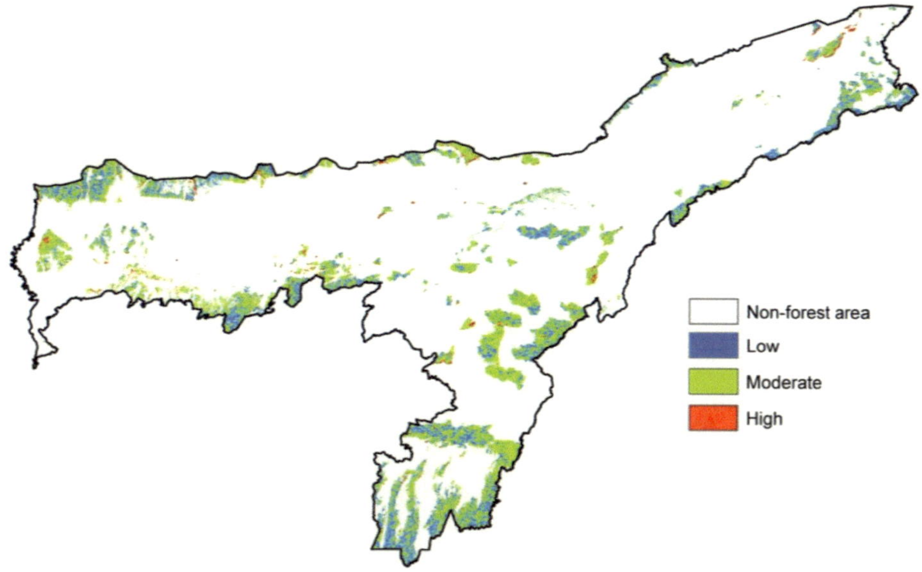

Figure 8.
Spatial distribution of non-instantaneous disturbance categories.

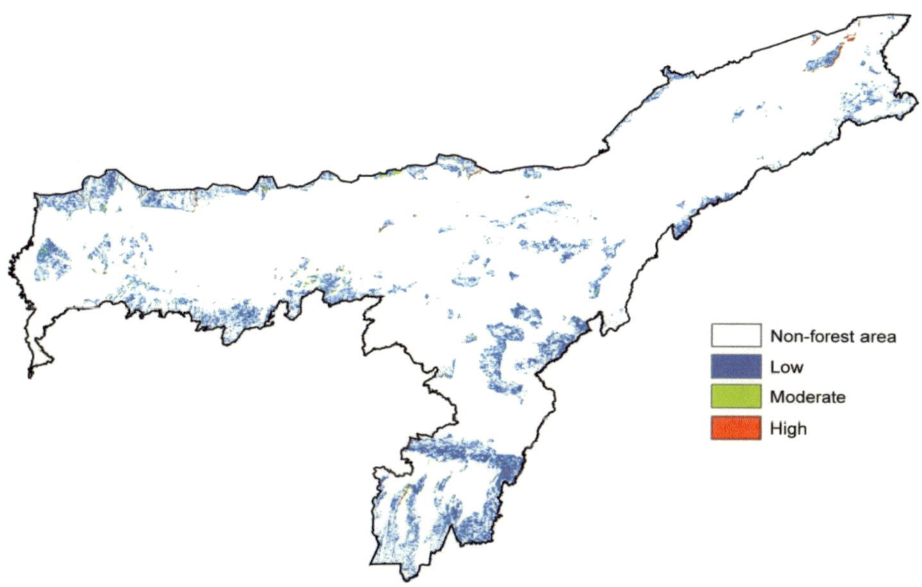

Figure 9.
Spatial distribution of instantaneous disturbance categories.

change of MGDI, both the disturbance types are classified into low, medium and high category to spatially depict the disturbance regime. It was seen that most of the non-instantaneous disturbed area falling under "low" category, followed by "moderate". Patches of "highly disturbed" areas are observed in north of Tinsukia district which falls under the flood plain of Brahmaputra River. Scattered patches of "highly disturbed" areas also noticed along the northern boundary of the Assam state adjoining to Bhutan, Kokrajhar and at the border of Karbi Anglong (East) and Golaghat districts. Under instantaneous disturbance regime most of the forest areas are falling under "low" disturbance category presumably due to high threshold value. Notably the forests of north Tinsukia district falls under "highly disturbed" category in both the instantaneous and non-instantaneous disturbance.

6. Conclusions

The ecological disturbance regime of the Assam forest area was assessed using Global Disturbance Index derived from the time-series MODIS EVI and LST data. The % change in MGDI from its multi-year mean was found to be in good agreement with the flood as well as forest fire intensity. The thresholds for non-instantaneous disturbance, i.e. flood, was found to be lower than the instantaneous disturbance, i.e. forest fire. The time-series disturbed area maps were able to capture the spatio-temporal dynamics of the disturbance regimes. The high disturbed area due to flood were in good agreement with the high rainfall year. The temporal profiles of the forest specific disturbed area could able to distinguish the major disturbed years. The disturbed prone area maps were able to classify the Assam forest areas into three major classes, which can be further utilized for the better management of the forest areas. The main assumption of the study was that the disturbances were created due to two natural hazards, like flood and forest. However, disturbances can be caused by disease/pests, anthropogenic interference, climate change etc. which needs to be examined. Hence, future study can be adopted for estimating the disturbance regime using multiple factors with an intensive ground data support. In the present study, forest fire events was used for estimating the fire intensity, whereas with the aid of the forest fire extent the present methodology would have been more robust. Sensors with better spatial resolution would increase the estimation accuracy of the disturbed area for localized disturbance.

Acknowledgements

The authors are thankful to the Director, NRSC and Chief General Manager (Regional Centres), for their continuous support and encouragements during the investigation. We duly acknowledge the MODIS data site for time-series data supply and forest fire information. We also acknowledge the Disaster Management Support Service Group, NRSC, for providing the flood inundation maps.

Conflict of interest

The authors declare no conflict of interest.

Author details

Dibyendu Dutta[1*], Akanksha Balha[2], Prabir Kumar Das[1], Pragyan Jain[3], Libeesh Lukose[4] and Wasim Akram[5]

1 National Remote Sensing Centre (NRSC), New Town, Kolkata-700156, India

2 TERI School of Advanced Studies, New Delhi-110070, India

3 GGITS, Jabalpur – 482003, MP, India

4 Department of Geology and Geophysics, IIT, Kharagpur - 721302, India

5 Vidyasagar University, West Midnapur - 721102, West Bengal, India

*Address all correspondence to: ddisro@gmail.com

References

[1] Dudley N, Mansourian S, Stolton S, Suksuwan S. 2010. Do protected areas contribute to poverty reduction?. Biodiversity. 2010; 11: 5-7. DOI: 10.1080/14888386.2010.9712658.

[2] Hannah L, Midgley G, Andelman S, Araujo MB, Hughes G, Martiner-Meyer E, Pearson RG, Williams P. Protected area needs in a changing climate. Frontiers in Ecology and Environment. 2007; 5: 131-138. DOI: org/10.1890/1540-9295 (2007)5[131:PANIAC]2.0.CO;2.

[3] Amiro BD, Todd JB, Wotton BM, Logan KA, Flannigan MD, Stocks BJ. Direct carbon emissions from Canadian forest fires, 1959-1999. Canadian Journal of Forest Research, 2001; 31: 512-525.

[4] Canadell JG, Mooney HA, Baldocchi DD, Berry JA, Ehleringer JR, Field CB. Carbon metabolism of the terrestrial biosphere: A multi-technique approach for improving understanding. Ecosystems, 2000; 3: 115-130.

[5] Fraser RH, Li Z, Cihlar J. 2000. Hotspot and NDVI Differencing Synergy (HANDS): A new technique for burned area mapping over boreal forest. Remote Sensing of Environment, 2000; 74: 362-376.

[6] Kurz WA, Stinson G, Rampley GJ, Dymond CC, Neilson ET. 2008. Risk of natural disturbances makes future contribution of Canada's forests to the global carbon cycle highly uncertain. Proceedings of the National Academy of Sciences of the United States of America. 2008; 105: 1551-1555. DOI: 10.1073/pnas.0708133105.

[7] Running SW. 2006. Is global warming causing more, larger wildfires. Science. 2006; 313: 927-928.

[8] Van derWerf GR, Randerson JT, Collatz GJ, Giglio L, Kasibhatla PS, Arellano AF Jr. Continental-scale partitioning of fire emissions during the 1997 to 2001 El Niño/La Niña period. Science. 2004; 303: 73-76. DOI: 10.1126/science.1090753.

[9] Picket STA, White PS (Editors) The Ecology of Natural Disturbance as Patch Dynamics. London/Orlando, FL, USA: Academic Press; 1985. 472 p. ISBN 0-12-554520-7.

[10] Tilman D. The resource-ratio hypothesis of plant succession. American Naturalist. 1985; 125(6): 827-852.

[11] Alcaraz-Segur D, Paruelo JM, Cabello S 2009. Baseline characterization of major Iberian vegetation types based on the NDVI dynamics. Plant Ecology. 2009; 202: 13-29. DOI: 10.1007/s11258-008-9555-2.

[12] Kinyajui MJ. NDVI based vegetation monitoring in man forest complex, Kenya. African Journal of Ecology. 2011; 49: 165-174. DOI. org/10.1111/j.1365-2028.2010.01251.x.

[13] Kerr JT, Ostrovsky M. 2003. From space to species: ecological applications of remote sensing. Trends in Ecology and Evolution. 2003; 18, 6: 299-305.

[14] Mildrexler DJ, Zhao M, Running SW. Testing a MODIS Global Disturbance Index across North America. Remote Sensing of Environment. 2009; 113: 2103-2117. DOI:10.1016/j.rse.2009.05.016.

[15] Nemani RR, Running SW. Estimation of regional surface resistance to evapotranspiration from NDVI and Thermal-IR AVHRR data. Journal of Applied Meteorology. 1989; 28: 276-284. DOI: org/10.1175/1520-0450(1989)028<0276:EORSRT>2.0.CO;2.

[16] Dutta D, Das PK, Paul S, Sharma JR and Dadhwal VK. Assessment of ecological disturbance in the mangrove forest of Sundarbans caused by cyclones using MODIS time-series data (2001-2011), Natural Hazards. 2015; 79(2): 775-790. DOI 10.1007/s11069-015-1872-x.

Chapter 5

Delineation of Open-Pit Mining Boundaries on Multispectral Imagery

Ioannis Kotaridis and Maria Lazaridou

Abstract

During the last decades, monitoring the spatial growth of open-pit mining areas has become a common procedure in an effort to comprehend the influence that mining activities have on the adjacent land-use/land-cover types. Various case studies have been presented, focusing on land-cover mapping of complex mining landscapes. They highlight that a rapid as well as accurate approach is critical. This paper presents a methodological framework for a rapid delineation of open-pit mining area boundaries. For that purpose an Object-Based Image Analysis (OBIA) methodology is implemented. Sentinel-2 data were obtained and the Mean-Shift segmentation algorithm was employed. Among the many methods that have been presented in literature in order to evaluate the performance of an image segmentation, an unsupervised approach is carried out. A quantitative evaluation of segmentation accuracy leads to a more targeted selection of segmentation parameter values and as a consequence is of utmost importance. The proposed methodology was mainly conducted through python scripts and may constitute a guide for relevant studies.

Keywords: OBIA, image segmentation, lignite mine, open-pit mining, Sentinel-2

1. Introduction

1.1 Mining activity and remote sensing

Mining comprises the activity that includes the extraction of geological materials from earth with tunnels, shafts or pits. Mining and mines can be classified in several ways. According to materials commonly mined, three classes of mining can be distinguished: metallic, non-metallic and fuel minerals. Based on the nature of excavation, mineral extraction can be categorized into two classes: underground and surface mining. The latter includes open-pit mining (also known as open-cast mining) that is implemented to extract deep and massive deposits that are not covered by a thick overburden. Underground mining of such deposits would be disadvantageous, since the material is mainly close to the surface [1].

Greece has been commonly included in the top lignite producers in Europe [2]. Mining of fuel minerals constitutes a critical activity, since a large percentage of the country's energy needs is covered by a solid fuel, lignite. The first lignite mine in Greece appeared in 1873, whereas systematic exploitation commenced after

1950. Nowadays, the primary lignite extraction basins are located in Ptolemaida and Amyntaio. Lignite exploitation in Greece is conducted by surface mining and specifically open-pit mining [3].

Mining and specifically coal mining activities may cause severe environmental impacts [4]. Landscapes formed by mining activities are vulnerable to several geomorphic hazards, for instance, landslides and rockfalls [1]. The stability of excavations is a critical aspect of Greek lignite mines which become larger and deeper in comparison with those in the past. During the last few years, many events of severe deformations and disastrous slope failures occurred [2]. In addition, flood is a probable hazard, since water can enter pits and tunnels [1]. Surrounding areas are affected by mining with economic, environmental and social impacts [5].

Taking into consideration their synoptic coverage and multitemporal data acquisition capabilities, remote sensing methods have been widely implemented in applications related to mining activities. Availability of high spatial resolution data resulted in an increased interest of using satellite data to monitor surface mining activities [4]. Remote sensing offers a valuable tool for acquiring rigorous data, while decreases the cost of field surveys both in time and money [6].

Remote sensing applications related to mining activities include the following: mapping of the surface mineralogy, topography and related changes that are quite valuable throughout the operation and planning of mine, identifying and monitoring environmental effects and mapping surface movements of mine structures in order to monitor safety features [7]. Furthermore, the size and location of mine areas as well as land-cover changes due to mining can be extracted from satellite images [5]. Remote sensing can make mine planning procedures easier, enhance safety during and after mine operation and monitor environmental effect as well as rehabilitation [7].

1.2 Image segmentation in OBIA

In contrast with traditional pixel-based approaches, the primary methodological component in Object-Based Image Analysis (OBIA) is the image object [8, 9]. OBIA produces meaningful image objects only if the imagery is partitioned into similar or relatively similar areas. This requires a low value of internal heterogeneity regarding the parameter that is examined in comparison with its adjoining areas [8].

Image segmentation is the first but also fundamental procedure to produce the core elements of OBIA [10]. It is about the partitioning of an image into spatially adjoining and homogenous groups of pixels (segments) that constitute the foundation for further analysis [8, 11]. These regions have similar spatial and spectral features, which, if considered as meaningful, depict a real-world object [9, 12]. By implementing image segmentation, the level of detail is decreased to make image content more comprehensive by lessening image complexity [9]. By transitioning from pixel to image object-based framework, in an effort to follow the example of visual interpretation, better management of spatial information can be accomplished, thus a more beneficial integration with Geographic Information System (GIS) can be achieved [13].

During the last decades, several segmentation methods were matured and employed in remote sensing applications [10]. Commonly, segmentation methods are classified into three broad categories: pixel-based, edge-based and region-based methods [14]. The selection of segmentation method is substantially influenced by the objective of image analysis study and it is typically acknowledged that it does not exist a perfect algorithm that will demonstrate adequate results with every satellite image. It has to be mentioned that most segmentation methods do not instantly produce meaningful image objects. However, clusters are generated with generic labels, for example, region A, region B, etc. Then, these clusters have to be converted to meaningful image objects through a post-segmentation process [15].

A fairly demanding task in image segmentation procedure is the selection of segmentation parameters' values in order to generate segments that will comply with the needs of user and the purpose of study [10]. Since there is not a commonly accepted method to determine optimal segmentation parameters' values, image segmentation continues to be an interactive procedure that includes trial-and-error approaches.

A typical OBIA approach includes two main steps, image segmentation and object classification. On the other hand, there are studies that propose a methodology that includes only the step of image segmentation. It hast to be noted that the application objective is the definitive factor concerning the methodology implemented. This study does not follow the traditional OBIA approach.

1.3 Relevant studies

Several studies have utilized satellite data and remote sensing methods to investigate an issue related to mining activities. Monitoring and evaluating reclamation procedure in mining areas is a common application [16]. LaJeunesse Connette et al. [17] developed a methodology to detect mining areas and evaluate mining expansion in Myanmar. For this reason they used data free of charge and open-source software. Likewise, Li et al. [18] employed multitemporal Landsat data to monitor the expansion of coal mining activity. Demirel et al. [6] proposed a methodology for detecting land use changes in surface coal mines with the use of multi-temporal high-resolution satellite data. Similarly, Guan et al. [19] investigated land use changes in a surface coal mine area located in the northeast China. In addition, Latifovic et al. [20] presented a methodology for land-cover change evaluation in the Athabasca Oil Sands region, northeast Alberta, Canada. For this purpose Landsat data were obtained. Maxwell et al. [21] combined very high resolution imagery and LIDAR data for mapping land-cover of a surface coal mine area in the southern coalfields of West Virginia, USA. Demirel et al. [22] investigated the potential implementation of a machine learning classifier (Support Vector Machines) for classifying high spatial resolution multispectral data of an open-cast mine area. Lechner et al. [23] carried out a spatial assessment of mine disturbance and rehabilitation of an open-pit mining study area. Townsend et al. [24] presented a methodology for quantifying land-use and land-cover change patterns due to surface mining and reclamation in the Central Appalachian Mountain region of the Eastern U.S., during a 30-year timeframe.

1.4 Scope of the study

The primary objective of the present work is to provide an object-based methodology for rapid detection and delineation of an open-pit mining area boundaries located nearby Amyntaio town, in northwestern Greece. Since image segmentation quality is a critical part in our analysis, an unsupervised evaluation of image segmentation performance was conducted, quantifying the internal homogeneity of segments and between segment separability.

2. Study area and data

The study area that was selected for this paper covers the Public Power Corporation (PPC) SA Amyntaio lignite mine. It has an extended mining history

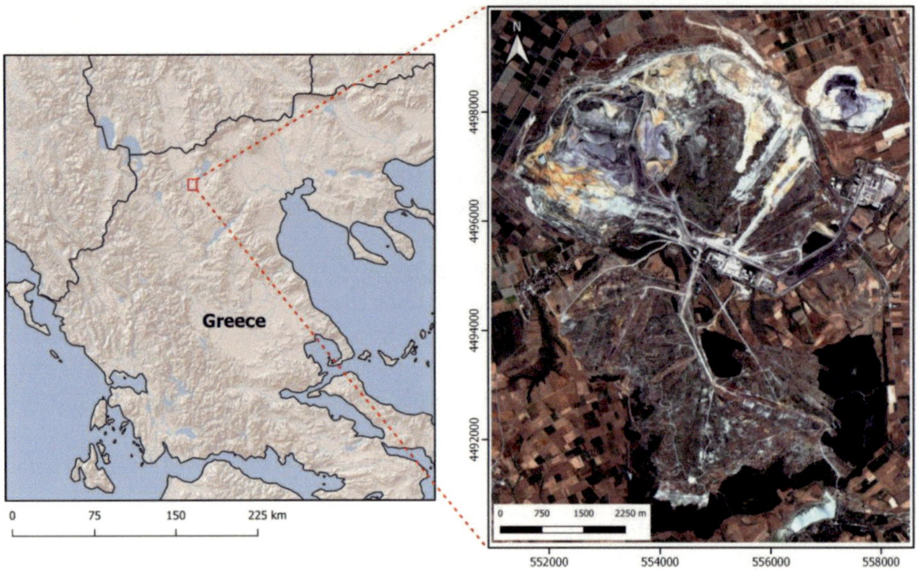

Figure 1.
Study area located in Greece (left) and the subset of Sentinel-2A imagery (2020) (right).

that includes opencast mining with depths that reach 180 m, since the mid-1980s and is considered a critical mine for energy production in Greece. It is located in the north west part of mainland Greece. A subset was extracted from the main scene for analysis in order to include the mining operations as it appears in **Figure 1**, that is presented below.

In this paper Sentinel-2A, Level 2A (Bottom-Of-Atmosphere) corrected reflectance imagery was obtained. The scene acquisition date is 30 June 2020 (Tile T34TEK). The criteria for the selection of scene were limited cloud coverage and high quality.

3. Methodological procedure

The methodology implemented in order to delineate the boundaries of this open-pit mining area is presented in **Figure 2**.

3.1 Tools

Orfeo Toolbox (OTB) was used for digital processing of the imagery. It is an open-source project that supports processing of remote sensing data including high resolution optical, multispectral and radar images [25]. The algorithms utilized for the purpose of this study were accessed from Python through the otbApplication module. Spatial analysis procedures were carried out in QGIS, a free and open-source Geographic Information System that supports the creation, editing visualization and publication of geospatial data [26].

3.2 Initial processing of data

Initial processing of Sentinel-2A imagery includes resampling the 20 m bands to 10 m, clipping the scene to the boundaries of Area Of Interest (AOI) and concatenating the spectral bands to produce a single stacked image.

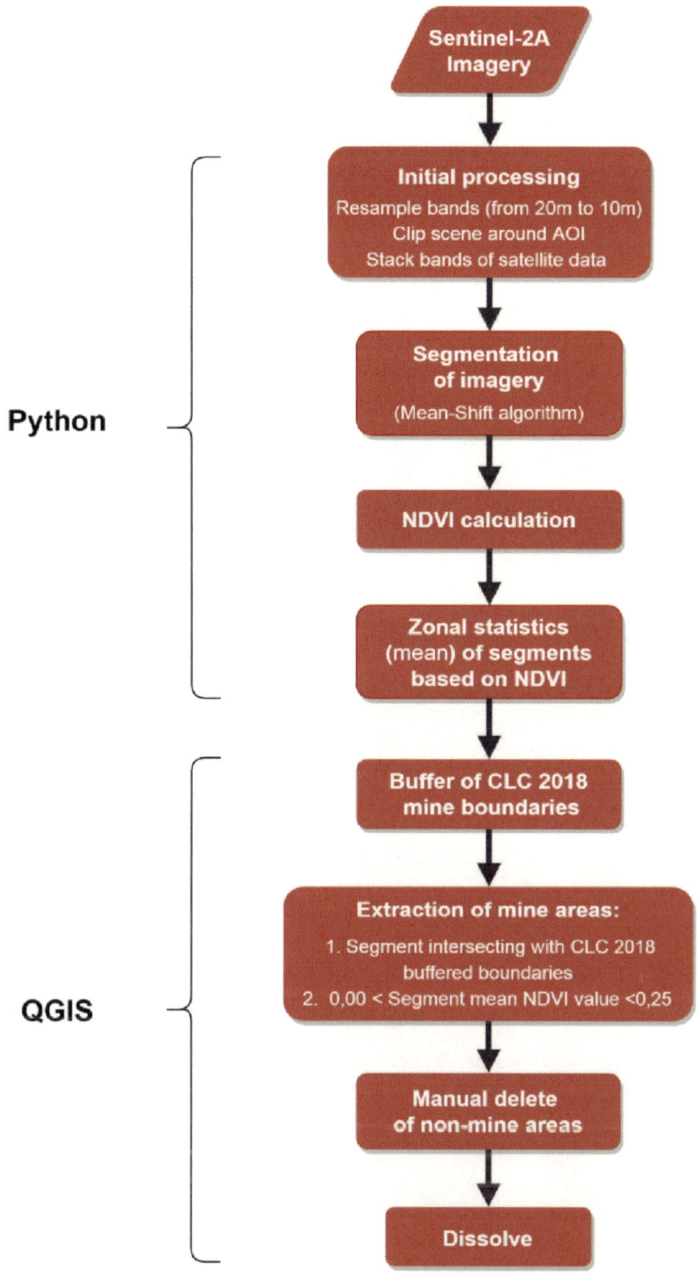

Figure 2.
Methodological framework.

3.3 Image segmentation

The stacked image was used as input for image segmentation. For the purpose of this study, Mean-Shift segmentation algorithm was implemented. It is a non-parametric clustering approach that is widely utilized in image analysis [27]. Mean-Shift segmentation algorithm has depicted adequate results regarding object extraction [28, 29]. It can handle different remote sensing satellite data, for instance, medium or high spatial resolution images. Critical factors that are related to its popularity are the simplicity of filtering step, the multivariate nature and the existence of various implementations [30]. The vector output of segmentation and the selected parameter values of Mean-Shift algorithm are presented in **Figure 3**.

As shown in **Figure 3**, range radius and minimum region size parameter values were increased drastically compared to default values. This is necessary in order not to confuse mine areas with different land-cover types and not to have a large number of segments, so that they are manageable.

3.4 Evaluation of segmentation

A qualitative evaluation of segmentation output is commonly implemented through visual assessment [31]. This is a rather subjective mean of segmentation accuracy evaluation. Conversely, several supervised and unsupervised approaches

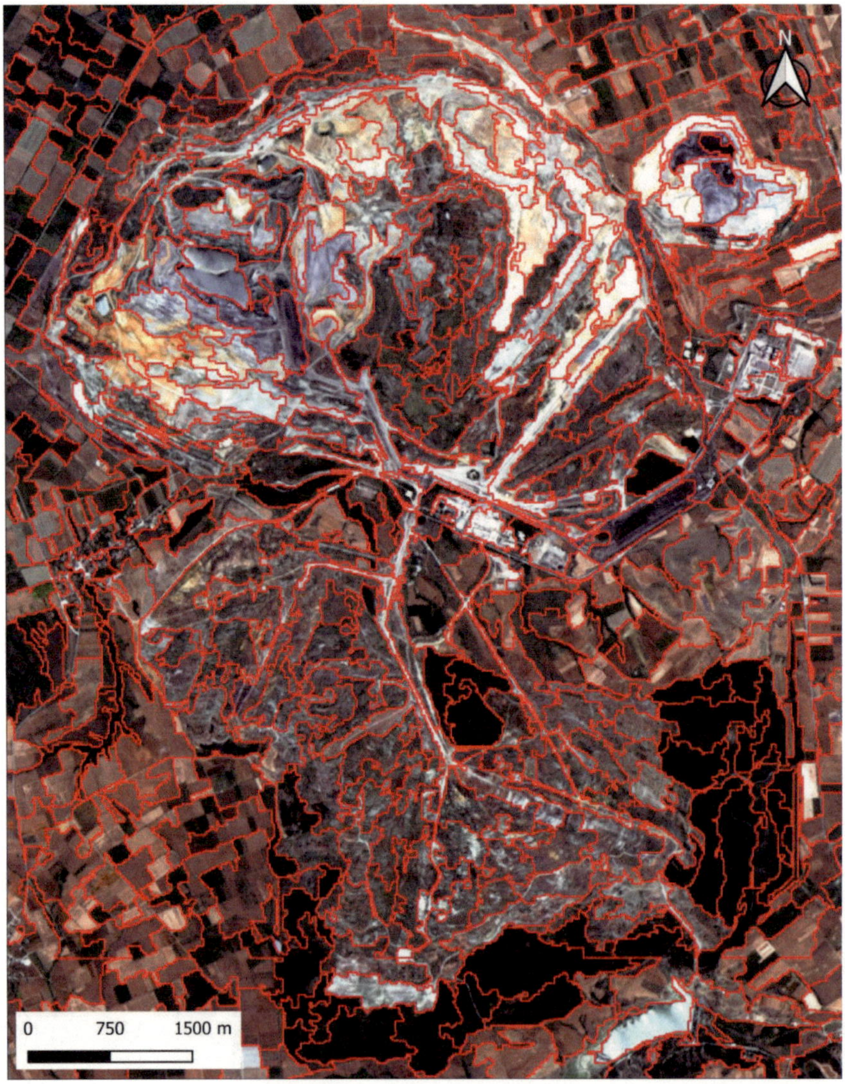

Figure 3.
Segmentation results. The boundaries of segments are symbolized with red color (spatial radius: 4, range radius: 160, minimum region size: 700).

have been presented in order assess image segmentation accuracy. Supervised methods typically compare segmentation output with a reference layer and measure the overlapping area [32]. Unsupervised approaches measure particular features of segments, for example, spectral homogeneity and between object heterogeneity [33]. However, there is not a standard methodology [32].

For the purpose of this study, an unsupervised approach was selected. In specific, the objective function proposed by Espindola et al. [33] was calculated to evaluate the quality of image segmentation results. This function consists of a measure of intrasegment homogeneity and one of intersegment heterogeneity. The first part is intrasegment variance of segments, a weighted average, where the area of each segment represents the corresponding weight. Thus, probable variabilities produced by smaller segments are eliminated. Furthermore, in order to evaluate intersegment heterogeneity, the function employs Moran's I autocorrelation index [34] that measures the spatial association as derived from the total of segments. Moran's I reflects how, on average, mean values of each segment vary from mean values of its adjacent segments. Small values of Moran's I suggest low spatial autocorrelation, hence the adjacent regions are statistically different. This denotes large intersegment heterogeneity. In other words, image segmentation produces segments with discrete boundaries. Employing spatial autocorrelation for evaluating image segmentation quality is especially suitable for region growing algorithms that generate closed polygons [33].

An adequate selection of parameters' values incorporates low intersegment Moran's I index with low intrasegment variance. The proposed function from Espindola [33] adds the normalized values of variance and autocorrelation measures. The objective function and its components were computed for each spectral band of Sentinel-2 imagery. Following, the value of objective function for the entire imagery was calculated by averaging the values of each spectral band. The results are presented in **Table 1**.

As shown in **Table 1**, the mean normalized value of variance slightly changes for different parameters' values and the lowest values corresponds to the lowest values of range radius and minimum region size, as expected. Moran's I index value is decreasing when range radius value and minimum region size are increasing, which means that segments get larger in size but also fewer in number. The selected Mean-Shift parameters' values for this specific study area

Mean-Shift parameters' values (Spatial radius/Range radius/Minimum region size)	Variance	Moran's I index	Objective function
5/15/100	0.53	0.57	1.10
4/80/700	0.59	0.38	0.97
4/120/700	0.60	0.35	0.95
4/160/700	0.60	0.34	0.95
4/200/700	0.60	0.38	0.98
4/240/700	0.59	0.36	0.95
4/200/1500	0.61	0.27	0.89
4/300/2000	0.59	0.28	0.87

Table 1.
The values of variance, Moran's I index and objective function for specific Mean-Shift parameters' values.

(4/160/700) correspond to relatively low Moran's I index value which denotes that neighboring segments are statistically discrete.

3.5 NDVI calculation

Normalized Difference Vegetation Index (NDVI) was computed among several spectral indices. The relevant bands for NDVI are Red and NIR. NDVI is a simple but also undoubtedly effective and extensively implemented index for quantifying green vegetation. NDVI values range from −1 to +1. Negative values suggest the existence of water bodies. Values close to zero (−0.1–0.1) typically correspond to barren land. Values above 0.1 commonly indicate the existence of green vegetation [35]. NDVI of the study area is presented in **Figure 4**.

As shown in **Figure 4**, values from 0 to 0.2 clearly indicate the existence of mine areas, as it can be visually recognized from the natural color image. This

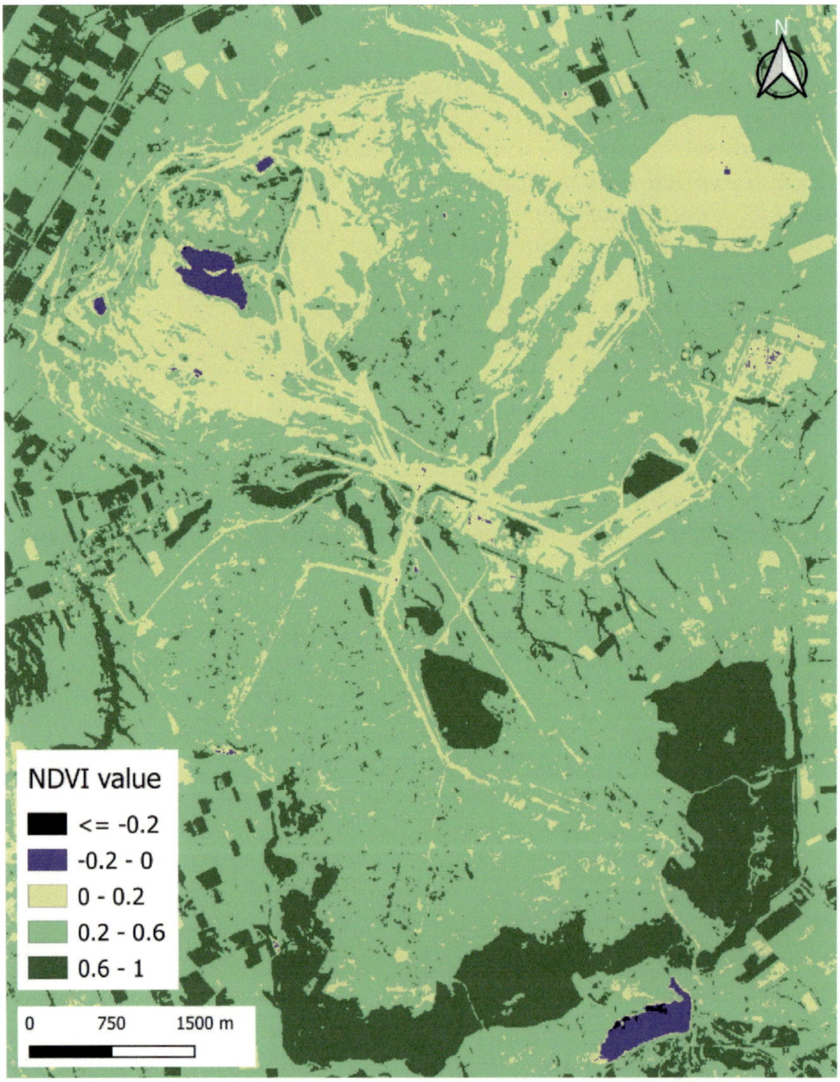

Figure 4.
NDVI of the imagery.

observation lead to the exploitation of this specific feature to extract the boundaries of mine areas.

3.6 Zonal statistics

Zonal statistics of segments were computed from NDVI raster layer. In specific, min, max, standard deviation and mean statistics were calculated over each segment. Following, they were exported in a vector layer (shapefile). It was ascertained that mean value statistic comprise an ideal indicator to identify mine areas. Mean value of NDVI for each segment is presented in **Figure 5** superimposed on the natural color image of the study area.

3.7 Delineation of mine area

Following the identification of mine areas, isolation of these areas is the next step. Since some areas outside mines share the same mean NDVI values with mines,

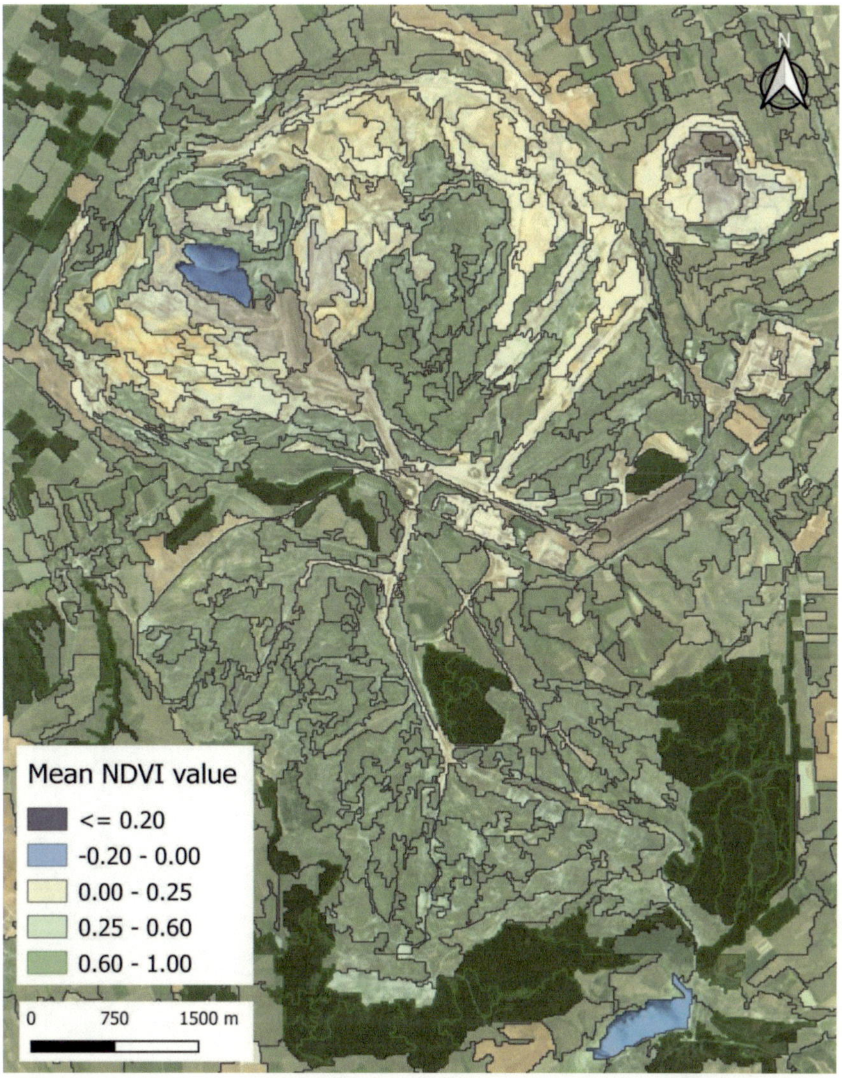

Figure 5.
Mean NDVI value of segments.

further processing of the vector layer is required. A dataset with rough boundaries of mine areas forms an auxiliary layer that can help to remove non-mine areas. In specific, Corine Land Cover (CLC) 2018 polygon of mine class was employed. A buffer area was computed using a fixed distance of 500 m for two main reasons. Minimum mapping unit of CLC datasets is 25 hectares, thus it is not appropriate for the scale of the analysis of this study and cannot be used unchanged. In addition, the reference year of satellite data used for the production of the latest CLC status layer is 2018, while the reference year of the imagery used in this paper is 2020. Several changes regarding mine boundaries occurred during this timeframe. CLC 2018 mine polygon and 500 m buffered polygon are presented in **Figure 6**.

A segment, in order to be characterized as mine area has to satisfy two conditions. It has to intersect with CLC 2018 buffered boundaries and its mean NDVI value has to be in the range of 0.00 to 0.25. Through this approach, segments were filtered and non-mine areas (polygons outside buffer zone) were erased. Furthermore, a manual more precise removal of non-mine areas was carried out to the remaining segments. The final step includes the implementation of dissolve algorithm in order to dissolve adjacent segments that share a common boundary.

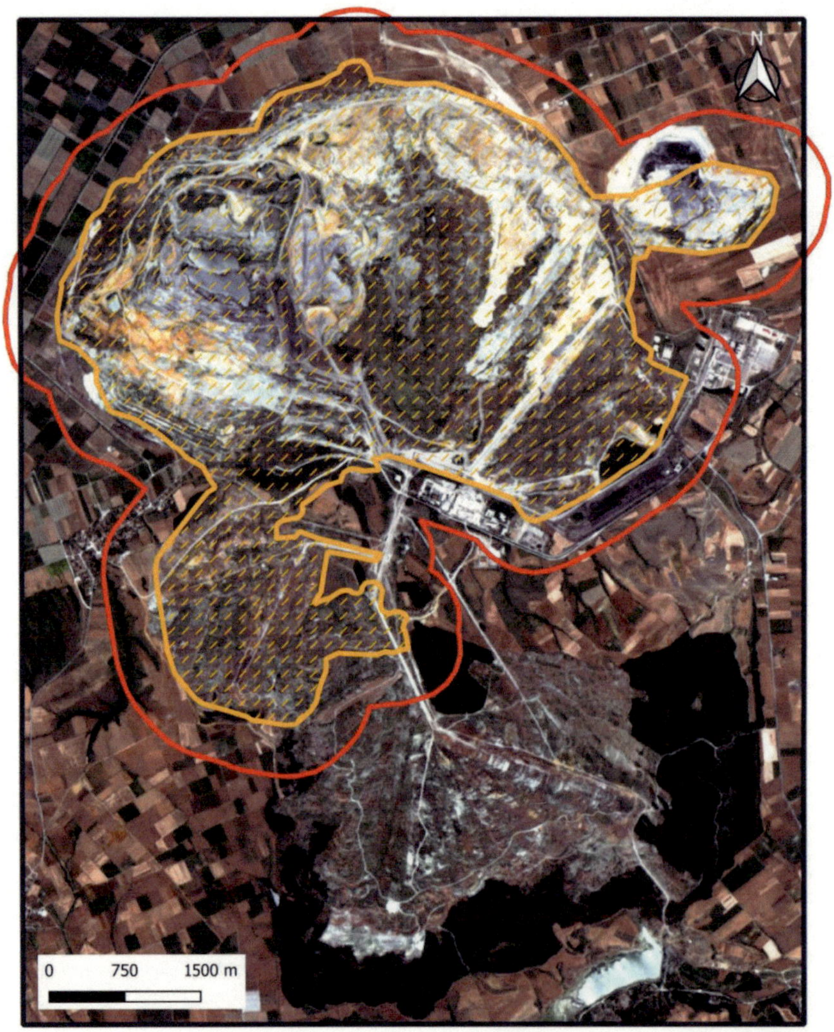

Figure 6.
CLC 2018 mine polygon (in orange color) and 500 m buffered polygon (in red color).

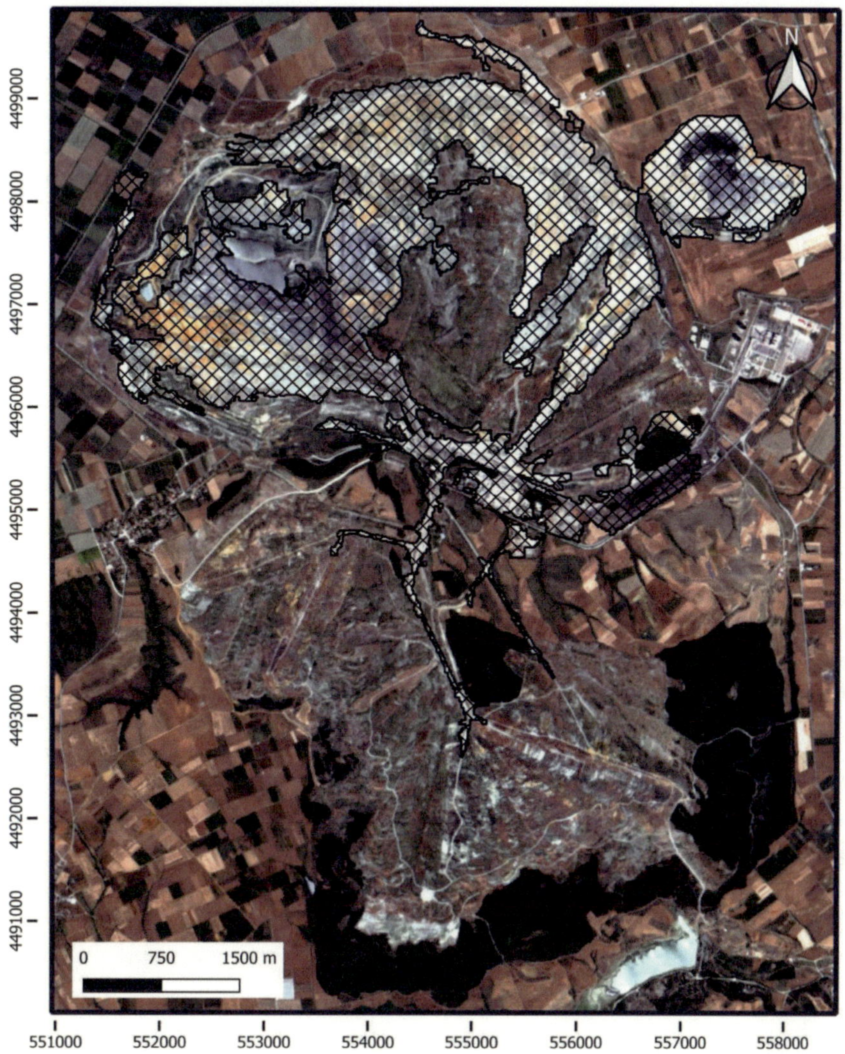

Figure 7.
Mine area in crosshatch pattern.

In this way, several segments were converted to a single meaningful image object (the mine area), that is demonstrated in **Figure 7**.

4. Discussion

An effective way of accomplishing adequate environmental management of mining areas requires the integration of remote sensing methods and Geographic Information Systems. Remote sensing provides image analysis fundamentals while a Geographic Information System offers spatial data analysis and geo-visualization tools. If these are exploited in a proper way, then continuous monitoring of mining activity can lead to efficient reclamation. In addition, freely-available data and open-source software drastically facilitates the efforts in this direction. This study utilized both of them in an effort to develop a comprehensive and at the same time rapid methodology for identifying mining areas and precisely delineating their boundaries. Of course, this approach can be beneficial for a multitemporal analysis in order to evaluate mining expansion.

The implemented approach for image segmentation evaluation that is demonstrated in this study does not require ground truth data, since it is an unsupervised method that is characterized by two features included in the following. Each segment should be internally homogeneous (weighted variance metric) and at the same time discrete from its adjacent segments (Moran's I spatial autocorrelation index). These two indicators are calculated for each spectral band and then combined into a global evaluation metric, the objective function. The main advantage of this approach is its robustness, since it exploits well-established statistical methods. However, since it is a global evaluation metric, it may not perform well when two segmentation results depict very similar performance but have dissimilar local error distributions. An approach that is capable of quantifying both locally and globally segmentation performance may be more suitable for the aforementioned situation.

5. Conclusion

In this study a methodology for rapid identification of mines and precise delineation of theirs boundaries is presented, with the use of both freely-available data and open-source software. For this reason a cloudless Sentinel-2A imagery was obtained covering the area of interest. Following the initial processing steps, image segmentation was carried out using Mean-Shift algorithm and an unsupervised segmentation evaluation metric was calculated for different parameters' values. It is combined by an autocorrelation index that identifies separability between segments and variance, an indicator that depicts the global homogeneity of segments. Then, NDVI and its mean values for each segment were computed. Finally, the mine area was extracted by implementing some spatial analysis tools including dissolve algorithm in order to aggregate segments that share a common boundary.

Author details

Ioannis Kotaridis* and Maria Lazaridou
Laboratory of Photogrammetry - Remote Sensing, Faculty of Engineering, School of Civil Engineering, Aristotle University of Thessaloniki, Thessaloniki, Greece

*Address all correspondence to: iskotarid@civil.auth.gr

References

[1] Mossa J, James LA. 13.6 Impacts of Mining on Geomorphic Systems. In: Treatise on Geomorphology [Internet]. 2013. p. 74-95. Available from: https://linkinghub.elsevier.com/retrieve/pii/B9780123747396003444

[2] Zevgolis IE, Deliveris AV, Koukouzas NC. Slope failure incidents and other stability concerns in surface lignite mines in Greece. Journal of Sustainable Mining. 2019 Nov;18(4):182-97.

[3] MiningGreece. [Internet]. Available from: https://www.miningreece.com/mining-greece/minerals/coal/ [Accessed: 2020-08-01]

[4] Rathore† CS, Wright R. Monitoring environmental impacts of surface coal mining. International Journal of Remote Sensing. 1993;14(6):1021-42.

[5] Werner TT, Mudd GM, Schipper AM, Huijbregts MAJ, Taneja L, Northey SA. Global-scale remote sensing of mine areas and analysis of factors explaining their extent. Global Environmental Change. 2020;60.

[6] Demirel N, Düzgün Ş, Emil MK. Landuse change detection in a surface coal mine area using multi-temporal high-resolution satellite images. International Journal of Mining, Reclamation and Environment. 2011;25(4):342-9.

[7] REMOTE SENSING FOR THE MINING INDUSTRY. [Internet]. 2018. Available from: https://www.oulu.fi/sites/default/files/36/RESEM_EOReview.pdf

[8] Blaschke T, Burnett C, Pekkarinen A. Image Segmentation Methods for Object-based Analysis and Classification. In: Jong SMD, Meer FDV, editors. Remote Sensing Image Analysis: Including the Spatial Domain. Dordrecht: Springer; 2004. p. 211-36.

[9] Lang S. Object-based image analysis for remote sensing applications: modeling reality – dealing with complexity. In: Blaschke T, Lang S, Hay GJ, editors. Object-Based Image Analysis. Berlin, Heidelberg: Springer Berlin Heidelberg; 2008. p. 3-27. (Lecture Notes in Geoinformation and Cartography). Available from: http://link.springer.com/10.1007/978-3-540-77058-9_1

[10] Cheng G, Han J. A survey on object detection in optical remote sensing images. ISPRS Journal of Photogrammetry and Remote Sensing. 2016;117:11-28.

[11] Nussbaum S, Menz G. Object-based image analysis and treaty verification: new approaches in remote sensing - applied to nuclear facilities in Iran. New York, NY: Springer; 2008. 170 p.

[12] Navulur K. Multispectral Image Analysis Using the Object-Oriented Paradigm [Internet]. CRC Press; 2006. Available from: https://www.taylorfrancis.com/books/9781420043075

[13] Castilla G, Hay GJ. Image objects and geographic objects. In: Blaschke T, Lang S, Hay GJ, editors. Object-Based Image Analysis. Berlin, Heidelberg: Springer Berlin Heidelberg; 2008. p. 91-110. (Lecture Notes in Geoinformation and Cartography). Available from: http://link.springer.com/10.1007/978-3-540-77058-9_5

[14] Blaschke T, Hay GJ, Kelly M, Lang S, Hofmann P, Addink E, et al. Geographic Object-Based Image Analysis – Towards a new paradigm. ISPRS Journal of Photogrammetry and Remote Sensing. 2014;87:180-91.

[15] Thenkabail PS, editor. Remotely Sensed Data Characterization, Classification, and Accuracies

[Internet]. Boca Raton, Florida: CRC Press; 2015. Available from: https://www.taylorfrancis.com/books/9781482217872

[16] Karan SK, Samadder SR, Maiti SK. Assessment of the capability of remote sensing and GIS techniques for monitoring reclamation success in coal mine degraded lands. Journal of Environmental Management. 2016;182:272-83.

[17] LaJeunesse Connette K, Connette G, Bernd A, Phyo P, Aung K, Tun Y, et al. Assessment of Mining Extent and Expansion in Myanmar Based on Freely-Available Satellite Imagery. Remote Sensing. 2016;8(11):912.

[18] Li N, Yan CZ, Xie JL. Remote sensing monitoring recent rapid increase of coal mining activity of an important energy base in northern China, a case study of Mu Us Sandy Land. Resources, Conservation and Recycling. 2015 Jan;94:129-35.

[19] Guan C, Zhang B, Li J, Zhao J. Temporal and spatial changes of land use and landscape in a coal mining area in Xilingol grassland. IOP Conf Ser: Earth Environ Sci. 2017;52:012052.

[20] Latifovic R, Fytas K, Chen J, Paraszczak J. Assessing land cover change resulting from large surface mining development. International Journal of Applied Earth Observation and Geoinformation. 2005;7(1):29-48.

[21] Maxwell AE, Warner TA, Strager MP, Pal M. Combining RapidEye Satellite Imagery and Lidar for Mapping of Mining and Mine Reclamation. photogramm eng remote sensing. 2014;80(2):179-89.

[22] Demirel N, Emil MK, Duzgun HS. Surface coal mine area monitoring using multi-temporal high-resolution satellite imagery. International Journal of Coal Geology. 2011;86(1):3-11.

[23] Lechner AM, Kassulke O, Unger C. Spatial assessment of open cut coal mining progressive rehabilitation to support the monitoring of rehabilitation liabilities. Resources Policy. 2016;50:234-43.

[24] Townsend PA, Helmers DP, Kingdon CC, McNeil BE, de Beurs KM, Eshleman KN. Changes in the extent of surface mining and reclamation in the Central Appalachians detected using a 1976-2006 Landsat time series. Remote Sensing of Environment. 2009;113(1):62-72.

[25] Orfeo Toolbox. Open Source processing of remote sensing images [Internet]. 2020. Available from: https://www.orfeo-toolbox.org/ [Accessed: 2020-08-01]

[26] QGIS. A Free and Open Source Geographic Information System [Internet]. 2020. Available from: https://qgis.org/en/site/ [Accessed: 2020-08-01]

[27] Remote Sensing for Forest Cover Change Detection. Module 3: Introduction to QGIS and Land Cover Classification. [Internet]. 2016. Available from: https://servirglobal.net/Portals/0/Documents/Articles/ChangeDetectionTraining/Module3_LC_Classification_Accuracy_Assessment.pdf [Accessed: 2020-08-01]

[28] Chehata N, Orny C, Boukir S, Guyon D, Wigneron JP. Object-based change detection in wind storm-damaged forest using high-resolution multispectral images. International Journal of Remote Sensing. 2014;35(13):4758-77.

[29] Huang F, Chen Y, Li L, Zhou J, Tao J, Tan X, et al. Implementation of the parallel mean shift-based image segmentation algorithm on a GPU

cluster. International Journal of Digital Earth. 2019;12(3):328-53.

[30] Michel J, Youssefi D, Grizonnet M. Stable Mean-Shift Algorithm and Its Application to the Segmentation of Arbitrarily Large Remote Sensing Images. IEEE Trans Geosci Remote Sensing. 2015;53(2):952-64.

[31] Hurskainen P, Adhikari H, Siljander M, Pellikka PKE, Hemp A. Auxiliary datasets improve accuracy of object-based land use/land cover classification in heterogeneous savanna landscapes. Remote Sensing of Environment. 2019;233:111354.

[32] Costa H, Foody GM, Boyd DS. Supervised methods of image segmentation accuracy assessment in land cover mapping. Remote Sensing of Environment. 2018;205:338-51.

[33] Espindola GM, Camara G, Reis IA, Bins LS, Monteiro AM. Parameter selection for region-growing image segmentation algorithms using spatial autocorrelation. International Journal of Remote Sensing. 2006;27(14):3035-40.

[34] Fotheringham AS, Brunsdon C, Charlton M. Quantitative geography: perspectives on spatial data analysis. Sage; 2000.

[35] Custom-scripts. Normalized difference vegetation index [Internet]. 2017. Available from: https://custom-scripts.sentinel-hub.com/sentinel-2/ndvi/ [Accessed: 2020-08-01]

Chapter 6

Stereoscopic Precision of the Large Format Digital Cameras

Benjamin Arias-Perez

Abstract

Stereoscopic vision is fundamental in the task of photogrammetric restitution (stereo compilation) in which, by inserting a floating mark in the 3D observation of pairs of images, it is possible to draw the elements of the terrain in space and obtain cartography of a part of the land cover from aerial images. Initially with film photographs, which were later scanned, and finally with large format digital cameras that began in the 2000s, photogrammetry has undergone a series of technological revolutions up to the present time. In this chapter, after a brief exposition of the basic principles of photogrammetric restitution, a review of current large-format digital cameras and their main implications in restitution is made, which, despite the advances and other similar semi-automatic products (DTM, orthophoto) is still manual and must be operated by a person with the implications that this entails in stereoscopic vision.

Keywords: photogrammetry, aerial digital cameras, stereoscopic precision, Ground Sample Data, photogrammetric restitution

1. Introduction

Photogrammetry (the art and science of determining the position and shape of objects from photographs [1]) has been used since the early 20th century as an efficient method of generating mapping of large areas of the territory, from images obtained with cameras on board an aircraft.

The "General Method of Photogrammetry" describes the stereoscopic processing of images: acquisition of a pair of images that verify artificial stereoscopy conditions; orientation of the images to each other; and virtual three-dimensional exploration of the stereoscopic space generated and cartographic capture of points:

- Acquisition of a pair of images that verify artificial stereoscopy conditions. The process of artificial stereoscopic vision is based on generating an image for the left eye and an image for the right eye (as in natural stereoscopic vision). In photogrammetry, images are separated from each other by a certain distance (called base, b) and the image axes are normal to this base and parallel to each other. This configuration is known as a normal case.

- Orientation of the images to each other. This achieves a stereoscopic model that is also metric (relative to an external reference system).

IntechOpen

- Virtual three-dimensional exploration of the stereoscopic space generated and cartographic capture of points. In 1892 Stolze invented the floating mark, which allows metric three-dimensional exploration. Two marks located in photography paths are perceived as a single point located in space. If the observer can move the marks on the images while receiving a stereoscopic perception of them, they can "pose the floating mark" on the surface of the object. This way of obtaining 3D coordinates is known as photogrammetric restitution or stereo compilation (**Figure 1**).

The *XY* precision is directly proportional to the scale of the image, m_b, and the measurement precision of the image, σ_i:

$$\sigma_{xy} = \sigma_i * m_b \tag{1}$$

The precision of the measure on the image plane, σ_i usually ± 6 μm [1] can be expressed in terms of pixel size, *px*, as a fraction (*1/k*). This value *k* can be considered as an indicator of measurement precision in the image.

$$\sigma_i = \frac{\text{px}}{k} \Rightarrow \sigma_{\text{xy}} = \frac{\text{px}}{k} * m_b \tag{2}$$

Moreover, the product of pixel size for image scale provides the pixel size in the ground, *GSD (Ground Sample Distance)*:

$$\text{GSD} = \text{px} * m_b \Rightarrow \sigma_{\text{xy}} = \frac{\text{GSD}}{k} \tag{3}$$

Thus, the precision observed in *XY* can be expressed as a fraction of the *GSD*. Once the empirical planimetric standard deviation, S_{XY}, is obtained, the empirical measurement precision of the image, S_i is get. From S_i the value of *k* can be computed which is a good value of comparison between cameras.

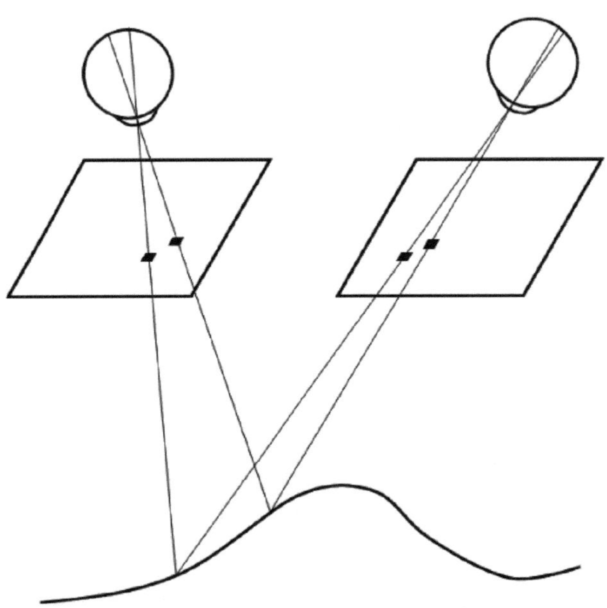

Figure 1.
Floating mark principle.

$$S_{xy} = S_i * m_b \Rightarrow S_i = \frac{S_{xy}}{m_b}$$
$$S_i = \frac{\text{px}}{k} \Rightarrow k = \frac{\text{px}}{S_i} \tag{4}$$

From this expression it follows that the higher k, the better precision.

It is important to note that σ expresses the theoretical precision while S expresses the empirical standard deviation which is determined from measurements.

The precision in Z, σ_Z, depends on the precision of measurement of the horizontal parallax, σ_{Px}, the image scale, m_b, and the ratio height/base, H/B [1]:

$$\sigma_z = \sigma_{\text{Px}} * m_b * \frac{H}{B} \tag{5}$$

The measurement precision of the horizontal parallax can be replaced by the measurement precision in the image plane, σ_i. The ratio height/base can be replaced by the ratio focal/photobase (c/b), then:

$$\sigma_z = \sigma_i * m_b * \frac{c}{b} \tag{6}$$

The precision of the measure in the image plane, σ_i, can be expressed in terms of pixel size as a fraction of it. In this case, it is assigned a value of *1/k*:

$$\sigma_i = \frac{\text{px}}{k}$$
$$\sigma_z = \frac{\text{px}}{k} * m_b * \frac{c}{b} \tag{7}$$

Moreover, the product of pixel size for image scale provides the pixel size in the ground, *GSD*:

$$\text{GSD} = \text{px} * m_b$$
$$\sigma_z = \frac{\text{GSD}}{k} * \frac{c}{b} \tag{8}$$

As can be seen, precision in Z can also be expressed in terms of the *GSD*, the focal length and photobase. This is a function of longitudinal overlap, R_L, together with the image width:

$$b = (1 - R_L) * \text{width} \tag{9}$$

Camera	c (mm)	Width (mm)	b (R_L = 60%) (mm)	c/b
Analogue	150	220	88	1,70
DMC	120	95	38	3,16
UltraCamD	100	67,5	27	3,70
DMC III	92	56,9	22.8	4,04
UltraCam Eagle M3 f80	80	68	2722	2,94
UltraCam Eagle M3 f210	210	68	27.2	7,72

Table 1.
Ratios c/b *or various photogrammetric aerial cameras, calculated for a longitudinal overlap of 60%.*

The value c/b affects proportionally the Z precision, so that the higher the value of this ratio less precision in Z (**Table 1**).

2. Aerial digital cameras

2.1 Technological evolution: film to digital

Photogrammetry has undergone three digital revolutions. The first took place in the 1980s with the digitization of the mathematical model that resulted in analytical stereoplotters (process input, frames, remained analogue). Analogue stereoplotters that solved the mathematical model by mechanical analogy disappeared.

The second occurred in the 1990s with digitization, using powerful and accurate photogrammetric scanners, analogue images from the aircraft that resulted in digital stations. This revolution was possible when personal computers had sufficient capacity to efficiently handle digital images, and represented the extinction of analytical stereoplotters and, with them, the stereoplotters themselves; that is, of the specific physical machines used for photogrammetry.

The third revolution took place in the first decade of the 21st century thanks to the development of digital cameras that began to compete with the large format (230 x 230 mm) of analogue cameras. These cameras were already used in terrestrial photogrammetry, where the small and medium formats were enough to develop projects. The need to cover large areas of land in aerial photogrammetry made the large size of the camera the only possible solution between the two photogrammetric requirements: accuracy (requiring long focal length), and performance in object coverage (requiring either short focal or large focal planes). This phase involves the disappearance of analogue cameras and, consequently, films. Therefore it will also represent the disappearance of photogrammetric scanners (**Table 2**).

The main advantages of digital versus analogue images are [2]: the ability to establish an entirely digital workflow (suppressing the scanners), a considerable improvement in radiometric quality as well as the possibility of simultaneously acquiring panchromatic images in the different color bands and in the near infrared; as well as the ability to generate real-time mapping [3]. The color depth of digital cameras (12-bit) should allow flights to be carried out in poor lighting conditions [4].

A first approach to digital cameras for photogrammetric use allows them to be classified into two large groups: frame and pushbroom. The first can be classified according to the size of the sensor [5]: small format cameras (up to 16 megapixels); medium format cameras (from 16 to 50 megapixels); and large format cameras (50 megapixels or higher). More recently, medium format can be located at 80–100 megapixels [6], and large format larger than 100 megapixels.

	It consists of	Appears	Disappears
First revolution (80s)	Digitizing the mathematical model	Analytical stereoplotters	Analogue stereoplotters
Second revolution (90s)	Post-flight scanning of the image	Digital station Photogrammetric scanner	Analytical stereoplotters
Third revolution (00s)	In-flight scanning of the image	Digital aerial camera	Analogue aerial camera Aerial films Photogrammetric scanner

Table 2.
Evolution of digital photogrammetry.

2.2 Cameras in the photogrammetric mapping sector

The two large manufacturers of analogue cameras for aerial photogrammetry took the first steps towards digital cameras around 2000, each opting for a different technology. Leica, who manufactured the RC30 model, used a pushroom camera, ADS40, based on space sensors (HSRC, WAAC) from the German institute DLR. Meanwhile, Z/I Imaging moved from the RMK-TOP analogue model to the DMC modular camera in 2000 (synchronous mode).

Following the launch in 2000 of ADS40 and DMC, Vexcel, manufacturer of scanners for photogrammetric use, offered the UltraCamD digital camera to the market in 2003 (syntopic mode).

Currently there are also other large format cameras and even some medium format ones competing for the same sector, but it is not our intention in this chapter to give a review of the current offer of cameras for mapping, only to study the characteristics related to stereoscopic accuracy. To do this, the current models of Leica, Z/I Imaging (now both in Hexagon Group) and Vexcel are considered representative: DMC III and UltraCam Eagle M3. In addition, they allow working with interchangeable lenses, which adds versatility so as to be able to use the most appropriate focal length for each use: short for mapping purposes, and long for orthophotos generation.

The sensors of large format digital cameras are clearly smaller in size than conventional cameras (Width in **Table 3**). To find an easily interpretable equivalence, we could say that, in order for digital cameras to compete with analogue, it must be assumed that there is an equivalence between 20 μm, the size of the scanned pixel, and the 10 μm average pixel size in a digital camera (**Table 3**).

Initially, the equivalence between 20 μm scanned and 10 μm digital is assumed by the manufacturers of these cameras because they think that the digital pixel is of higher quality than the scanned pixel [7]. According to these same authors, different experiments indicate that automatic matching is 2,5 times better in the case of a digital image.

Angular resolution refers to the angle subtended by a pixel from the point of view. The initial digital cameras can be considered equivalent to the 15 μm scan; however, the current ones are much better (less than 10″).

Figure 2 provides a comparison of a reticle photographed and scanned at a resolution of 5, 10 and 15 μm (a, b and c) while on the right (d) the same grid obtained directly by a digital camera appears.

Camera	Width (mm)	Pixel size (μm)	Focal lenght (mm)	Angular resolution (″)	AGL (m) for GSD = 0,10 m
Analogue	220	20	150	28	750
Analogue	220	15	150	21	1000
DMC	168	12	120	21	1000
UltraCamD	103,5	9	100	19	1111
DMC III	56,9	3,9	92	9	2359
UltraCam Eagle M3 f80	68	4	80	10	2000
UltraCam Eagle M3 f210	68	4	210	4	5250

Table 3.
Data of large format digital cameras.

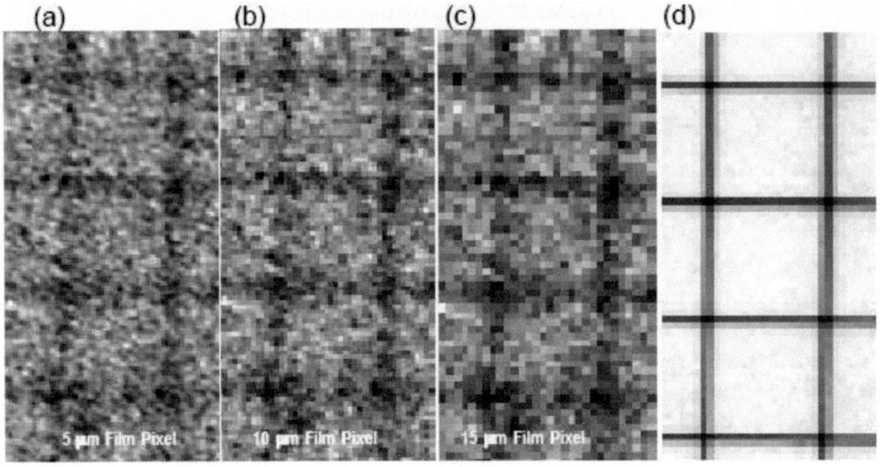

Figure 2.
A reticle photographed and scanned at a resolution of 5, 10 and 15 µm (a, b and c). The same grid obtained directly by a digital camera (d) [8].

3. Stereoscopic performance

Another issue in favor of digital cameras is their increased ability to get more images [7]. This gives them an advantage in terms of the possibility of achieving greater redundancy in the aerotriangulation process, and obtaining greater overlaps that decrease the overlap in the direction of flight.

The following sections provide a theoretical analysis of the stereoscopic performance of the DMC and UltraCam compared to the performance of analogue cameras.

3.1 Frame rate

Considering an airplane speed of 75 m/s and a *frame rate* of 1 second, the movement of the aircraft between two shots is 75 metros. The image size, in the case of the UltracamD camera, is 7.500 pixels in the direction of flight (*along track*). If the longitudinal overlap is 60%, this means that the base is 3.000 pixels. In this way, the minimum GSD with stereoscopic overlap that can be obtained with 60% longitudinal overlap is 25 mm.

$$\mathrm{GSD}_{\mathrm{minimum}} = \frac{\mathrm{Desplaz}}{\mathrm{Px} * (1 - R_L)} \tag{10}$$

Where *Desplaz* is the displacement of the aircraft between two shots (75 m), *Px is* the number of pixels of the sensor in the direction of flight (7.500), R_L is the longitudinal overlap (60%).

This means that images with GSD from 25 mm to 90% cannot be achieved, because the plane cannot fly that slowly (750 pixels * 25 mm = 18,75 m). These 25 mm imply a flight height 277,78 m:

$$H = \frac{c}{\mathrm{sx}} * \mathrm{GSD} \tag{11}$$

where f is the focal length (100 mm), sx the pixel size (9 µm) and GSD is Ground Sample Distance (25 mm).

The following **Tables 4–6** show the resulting minimum *stereoscopic* GSD sizes for different longitudinal overlaps, for *UltraCamD, UltraCamX,* and *DMC cameras.*

RL (%)	GSD min (m)	H (m)
60	0,025	277,778
70	0,033	366,667
80	0,050	555,56
90	0,10	1111,111

It has been considered an aircraft speed of 75 m/s and a frame rate of 1 second (base on the ground of 75 meters).

Table 4.
Minimum stereoscopic GSD sizes and their corresponding flight height (H) for the UltraCamD camera (Px = 7.500; sx = 9 μm; f = 100 mm), given the desired longitudinal overlap (RL).

RL (%)	GSD min (m)	H (m)
60	0,020	276,451
70	0,027	368,601
80	0,040	552,902
90	0,080	1105,803

It has been considered an aircraft speed of 75 m/s and a frame rate of 1 second (base on the ground of 75 meters).

Table 5.
Minimum stereoscopic GSD sizes and their corresponding flight height (H) for the UltraCamX camera (P = 9.420; sx = 7.2 μm; f = 100 mm), given the desired longitudinal overlap (R_L).

RL (%)	GSD min (m)	H (m)
60	0,049	488,281
70	0,065	651,042
80	0,098	976,563
90	0,195	1953,125

It has been considered an aircraft speed of 75 m/s and a frame rate of 1 second (base on the ground of 75 meters).

Table 6.
Minimum stereoscopic GSD sizes and their corresponding flight height (H) for the DMC camera (P = 7.680; sx = 12 μm; f = 120 mm), given the desired longitudinal overlap (RL).

3.2 Coverage

A fundamental part of any photogrammetric flight project is determining the direction and number of passes, number of total and past photographs, among other data. All this is determined from a series of initial conditions that establish the work area, which define the scale of the photograph to be obtained, etc...

First, a flight made with analogue camera is analyzed, for example, with the following characteristics:

- Scale of photography: 1:20.000
- Scanning photo size: 15 μm.
- Useful photo size: 220x220 mm.

This provides a GSD of:

$$15\mu m * 20.000 = 300 \text{ mm} \tag{12}$$

considering that each frame has a useful format of 220x220 mm, due to the space reserved in the frame for marginal information, the surface contained per frame is:

$$\begin{aligned} &220 \text{ mm} * 20.000 = 4.400 \text{ m} \\ &4.400 * 4.400 = 1.936 \text{ Ha} \end{aligned} \tag{13}$$

Now, with that same GSD, the resulting area for the *UltraCamD image* is:

$$\begin{aligned} &11.500 \ \text{ pixels} * 300 \ \text{ mm} = 3.450 \ \text{ m} \\ &7.500 \ \text{ pixels} * 300 \ \text{ mm} = 2.250 \ \text{ m} \\ &3.450 * 2.250 = 776,25 \text{Ha} \end{aligned} \tag{14}$$

Comparing both surfaces,

$$\frac{1.936}{776,25} = 2,49 \tag{15}$$

This means that approximately the 2.5-frame area of *UltraCamD is required* to cover the same surface as an analogue image. However, in the particular case of a frame, this is not true due to the different shapes of these (analog and rectangular square in digital camera). However, if the approach is generic, i.e., to cover a certain area for a project, that relationship can actually be valid.

3.3 Number of frames

Considering first longitudinally, and with the overlap of 60%, each new frame adds one side of 40% more to the strip. That is, the covered length, *L*, by *n* frames of a certain width, is determined by:

$$\begin{aligned} L &= \text{width} * \text{n} * 40\% \\ L &= \text{width} * \text{n} * (100 - R_L) \end{aligned} \tag{16}$$

To relate the number of photos to both flights to cover the same *length L*, with the same longitudinal overlap, this ratio is determined by the relationship between the widths of the frames, calculated above (Eqs (13) and (14)):

$$\frac{4.400}{2.250} = 1,956 \tag{17}$$

Similar reasoning can be followed for cross-sectional overlap:

$$L = \text{high} * \text{n} * (100 - R_T) \tag{18}$$

which in the example provides the following relationship:

$$\frac{4.400}{3.450} = 1,275 \tag{19}$$

If we now want to know the relationship between the total number of photographs between the two flights:

Camera	Useful width	Useful height	Coverage	Ratio
Analogue	220 mm	220 mm	1.936 Ha	—
UltraCamD	7.500 pixels	11.500 pixels	776,25 Ha	2,49
UltraCamX	9.420 pixels	14.430 pixels	1.223,38 Ha	1,58
DMC	7.680 pixels	13.824 pixels	955,51 Ha	2,03

Table 7.
Surfaces contained in a frame for different cameras, considering in all cases a GSD, where ratio expresses the ratio between the coverage of the analog camera and the coverage of the digital camera 300 mm.

$$1,956 * 1,275 = 2,49 \tag{20}$$

corresponding to the amount found in Eq. (15).

3.4 Conclusions on coverage

The relationship between the number of frames taken with analog camera and *UltraCamD* digital camera depends on the size of the pixel in the field, GSD. That is, by imposing a certain size of GSD, the relationship between the number of frames in both cameras is determined at the same time.

With the digital camera it is enough to multiply the number of pixels in height and width of the image by the GSD to obtain the actual magnitudes of the terrain to be covered with each frame In addition, it is clear that said GSD, the pixel size in the CCD, and the focal length will determine the flight height and scale of the photograph (**Table** 7).

While with the analog camera it is necessary to determine either the frame scale or the scanning pixel (variable depending on the scanner). Depending on one parameter, the other parameter is obtained. However, it is true that it is scanned at 15–20 μm., therefore, usually this pixel size determines the frame scale.

Acknowledgements

Department of Cartographic and Land Engineering, University of Salamanca.

Author details

Benjamin Arias-Perez
Department of Cartographic and Land Engineering, University of Salamanca, Ávila, Spain

*Address all correspondence to: benja@usal.es

References

[1] KRAUS, Karl; WALDHÄUSL, P. Photogrammetry Fundamentals and Standard Processes. vol. 1. Dümmler, Bonn,, 1993, vol. 14.

[2] Heipke, C., Jacobsen, K., Mills, J. Editorial Theme issue: "Digital aerial cameras". ISPRS Journal of Photogrammetry and Remote Sensing. 2006;60.6:361–362. DOI: 10.1016/j.isprsjprs.2006.06.004

[3] Tempelmann, U., Börner, A., Chaplin, B., Hinsken, L., Mykhalevych, B., Miller, S., Recke, U., Reulke, R., Uebbing, R. Photogrammetric Software for the LH Systems Airborne Digital Sensor. International Archives of Photogrammetry, Remote Sensing and Spatial Information Sciences. 2000;33.B2:552–559.

[4] Markelin, L.,Ahokas, E., Honkavaara, E., Kukko, A. Peltoniemi, J. Radiometric quality comparison of UltraCamD and analog camera. International Archives of Photogrammetry, Remote Sensing and Spatial Information Sciences. 2005;34.

[5] Petrie, G. Further Advances in Airborne Digital Imaging - Several New Imagers Introduced at ASPRS. GeoInformatic. 2006;9.8:16–23.

[6] RAIZMAN, Yuri. Productivity Analysis for Medium Format Mapping Cameras. Photogrammetric Engineering & Remote Sending. 2018;84.5:235–238.

[7] Leberl, F., Gruber, M., Ponticelli, M., Bernoegger, S., & Perko, R. The UltraCam large format aerial digital camera system. Proceedings of the American Society For Photogrammetry & Remote Sensing. 2003;sn:5–9.

[8] Perko, Roland; Gruber, Michael. Comparison of quality and information content of digital and film-based images. International Archives of Photogrammetry, Remote Sensing and Spatial Information Sciences. 2002;34.3/B:206–209.

Chapter 7

Remote Sensing Applications in Disease Mapping

Sabelo Nick Dlamini

Abstract

Disease mapping utilizes disease maps as visual representations of sophisticated geographic data that provide a general overview of the disease situation in a defined geographic area. Epidemiology is concerned with investigating the causes of diseases, and often, these causes vary in frequency and in space. This variation in space gave a niche to remote sensing to find its way into the public health domain as disease researchers sought to investigate the explaining environmental and climatic factors. Studies have demonstrated the potential offered by remote sensing application to disease mapping and epidemiology and to support surveillance and control efforts. We used some examples from a case study conducted in Eswatini in Southern Africa. Remote sensing imagery when combined with GIS spatial analyses techniques could support and guide existing disease surveillance and control programs at local, regional, and even continental scales. Researchers have also studied factors influencing the patterns and distributions of vector-borne diseases at a variety of landscape scales. However, successful application of remote sensing technology depends on the ability of nonexperts' remotely sensed data and end users to access, retrieve, and analyze the data captured from satellites. The exploration of some of the opportunities presented by remote sensing to disease mapping and epidemiology is still unfolding as new opportunities are being presented.

Keywords: disease mapping, epidemiology, geostatistics, remote sensing, GIS

1. Introduction

Remote sensing could be described as the science of scanning the earth using sensors onboard a satellite platform launched into space or high flying aircraft to obtain information and also monitor land use and land cover changes on the earth surface [1]. Often, the monitored land use and land cover changes emanate from human activities and their interaction with the environment. The observation of the earth by the orbiting satellites is done at different geographic scales and at different intervals or revisit periods, which are both widely referred to as spatial and temporal resolutions, respectively. Over the years, there had been noticeable improvements in the spatial and temporal resolutions which had been accompanied by an interesting visibility and readability of the captured images. In addition, the number of spectral bands used by the sensors to capture images had also increased, aiding in an appealing appearance to the human eye. Consequently, even nonexpert remote sensing users had been drawn into remote sensing by the *beautiful pictures* and availability of some of the end products of remote sensing. There had been, therefore, an increase in the application of remote sensing products by other disciplines and fields of study. In public health, remotely sensed (RS) data products had been widely adopted and used in disease mapping and epidemiology. We used some examples from a case study conducted in Eswatini in Southern Africa.

Disease mapping referrers to the use of disease maps as visual representations of sophisticated geographic data that provide a general birdeye overview of the disease situation in a defined geographic area. On the other hand, epidemiology is a branch of medicine that deals with the incidence, distribution, and strategies for disease containment and control as well as other factors relating to human and animal health. Epidemiology is concerned with investigating the cause of disease, and often, these causes vary in frequency and in space. This variation in space is what gave a niche to remote sensing to find its way into the public health domain as disease researchers sought to investigate the explaining environmental and climatic factors. For this reason, remote sensing which provides a birdeye overview of the earth surface had continued to be applied in disease mapping for rapid risk assessment and monitoring efforts. As a result, remote sensing products have for quite a reasonable while been prolifically applied and used in disease mapping and epidemiology. Products of remote sensing include various vegetation indices and environmental proxies which are derived from satellite images and used to elucidate land use and land cover changes as well as to approximate environmental and climatic conditions on the earth surface.

Vegetation indices are mathematical combinations of different spectral bands that are designed to numerically separate or stretch the pixel value of different features in an image [2, 3]. RS data products had been used in a number of disease mapping and epidemiological studies such as in risk mapping of malaria [4, 5], soil-transmitted helminths [6], schistosomiases, and prediction of high risk areas for leishmaniasis in Brazil. Previous work include incorporation of RS data in human health studies and spatial targeting of trachoma control in Southern Sudan [7] by developing a national risk map and mapping tsetse fly habitat suitability among others [8]. In addition, identification of environmental risk factors for cholera using satellite-derived remotely sensed data products had been undertaken by [9]. Determination of population living in a city using remotely sensed data products was carried out in a study by Karume et al. [10], whereby a GeoEye satellite image at 50-m resolution was used, and population of the city was obtained by taking the number of houses times an average number of habitants per house. To date, an inexhaustible number of exploratory research studies in disease mapping and epidemiology had been undertaken using remote sensing vegetation indices as environmental and climatic proxies in combination with various modeling approaches.

One of the main advantages of using RS data products in disease mapping and epidemiology is its near real-time availability for rapid assessment of at risk areas and prediction of disease distribution, especially in inaccessible areas that may also lack baseline data [11]. The increase in the launch of higher resolution satellites sensors and advances in data processing techniques have enabled a wider adoption of RS data [12]. In economically disadvantaged areas with poor ground measurement meteorological station networks, RS data had often been preferred and used as environmental and climatic proxies in disease risk mapping and prediction. As new sensors with better spatial and temporal resolutions become available, new opportunities had been presented and explored in the application of remote sensing products in disease mapping and epidemiology [13].

From the early generation of ecological studies that demonstrated the capability of RS data products in disease mapping [13–16], there had been a sustained proliferation of such studies in disease mapping and epidemiology. The application of geostatistical techniques to identify spatial heterogeneities in disease distributions, patterns, and trends as well as forecasting for epidemic preparedness planning had been demonstrated in studies by Chilès and Delfiner [17]; and Tran et al. [18] *inter alia*. Geostatistics is a branch of statistics that is used to analyze and predict the values associated with spatial and temporal phenomena [17]. It is often incorporated into modeling through the use of coordinates attached to the data that are being analyzed. An almost similar terminology also commonly used in such analysis is the term "biostatistics," which also refers to a similar approach except that the data being analyzed would involve biological data. Such models are then referred to as space–time models, especially due to the fact that they also include the observation dates of the mapped data. The theory behind incorporation of RS data in disease mapping and epidemiology was based on the field-observed association between environmental conditions and some of the disease-causing vectors [18–20], in particular how they vary in geographic space. For instance, some studies have demonstrated the association between radiation reflectance as measured by satellites and certain land cover types which have been used as environmental and climatic proxies for measurement of presence or absence of a disease and its vectors [21].

Incorporating remotely sensed data products in spatial modeling had been common in most aspects of geographic analyses. From the early usage of aerial photographs taken onboard aircraft to the launch of satellite-based sensors, remotely sensed data products have been in the forefront of scientific research on the land use and land cover changes of the earth. Many other disciplines such as public health and epidemiology have until recently overwhelmingly taken up remotely sensed data products and utilized them in disease mapping and in improvement of their understanding of environmentally driven diseases. The utilization of RS data products, especially in disease mapping, for instance, in mapping vector-borne diseases, had been prolific over the years. As mentioned earlier, products derived from remote sensing had been widely used in models as environmental proxies to analyze various behaviors of observed spatial phenomenon. Furthermore, environmental proxies had been incorporated as covariates in statistical models aimed at mapping, analyzing, and predicting spatial phenomenon relating to disease epidemiology.

In statistical models, disease analysis had been pursued by adjusting models with spatial covariates derived from remote sensing and regressed with any outcome of interest. In such cases, spatial regression models had often been parameterized using data variables derived from remotely sensed products. As a result, there had been an unprecedented upsurge in the utilization of environmental variables and proxies derived from remote sensing in statistical analysis models attempting to map diseases such as vector-borne diseases, including malaria, dengue fever, chikungunya fever, zika virus, and leishmaniasis *inter alia* [22]. Studies mapping the spatial distribution of such vector-borne diseases had often applied remotely sensed data products as environmental proxies by approximating either environmental conditions or land use and land cover features. When the spatial covariates had been used in such a manner in models, they would often be referred to as predictors or risk factors since they tend to covary with the severity of the disease. This association of diseases with their environment was first noted in some of the early ecological studies that demonstrated the capability of remote sensing products in disease mapping, including authors such as Beck et al. [15]; Hay [16], and Thomson et al. [14] among others. As an example, malaria had on numerous cases been referred to as an environmental disease, and describing environmental risk factors associated with this disease had driven the research in malaria risk mapping [23].

Whereas, the early uptake of remotely sensed products was initially limited by the slow processing power of first-generation computers as well as the limited and cumbersome storage facility required to store remotely sensed and geographic information system (GIS) data sets; the advent of fast processing computer power had made it possible to include remotely sensed data in mapping studies. In addition, storage facilities of geographic data sets had improved from the large and difficult-to-handle cassettes to small, easy-to-carry hard drives and universal serial bus (usb) including portable external hard drives. Furthermore, the open access to remote sensing end products even for civilian usage had stimulated the interest of the scientific community and facilitated the uptake and application of remote sensing in disease mapping and epidemiology.

In areas with limited ground-based environmental data observation stations, remotely sensed data products became the only option available for mapping disease risk distribution. Such inaccessible areas included those with rugged terrain, areas in armed conflict, and those with limited resources which would not be enough to undertake field-based studies. Also, in cases where for example, ground-based weather stations had been used, they had often been limited by data discrepancies emanating from human interference, human error, instrument jamming, and in some cases, power failure. This would often result in missing data as observation could not be completed on those days when the weather station failed. During the processing and analysis of such data, the missing values in the data would often require sophisticated methods of data imputation which would not escape criticism from the scientific peer review community. Therefore, remote sensing products had been preferred because of the reliability of availability often in near real-time compared to other data sources which may require field surveys to be undertaken and thus too expensive to be repeated.

2. Mapping environmental diseases in the twenty-first century

Most of the epidemiological studies that had mapped vector-borne diseases in the context of the environmental factors associated with those diseases [18–20] had based their assumptions on the established scientific evidence that those environmental factors were associated with the disease outcome of interest. To date an increasing number of disease mapping and epidemiology studies continue to use environmental and climatic data to map and predict disease distribution in defined geographic areas. Such studies would often be used to help guide and target the deployment of health interventions to those areas that had been identified to have high burden of the mapped disease. As the resolutions (both spatial and temporal) of remote sensing sensors had improved from the first generation of this technology, so has the interest and confidence in the use of their data products increased among the scientific community. As much as early studies were limited by computer processing power and storage, they were also limited by poor spectral bands of sensors which could not faithfully enhance delineation and demarcation of features. The spectral bands refer to the recorded wavelengths of the electromagnetic spectrum recorded by a sensor during image acquisition. Recently, sensor spectral bands have also improved and could now resolve and aid multicolor image display during feature analysis and identification. These advances in image processing, visualization and display have also supported and enabled the uptake and appreciation of research findings from mapping efforts as the resultant maps became more *beautiful* in addition to providing more information. **Figure 1** is an example of a high-resolution satellite imagery that had been used to develop a land cover classification map for part of Eswatini as shown in **Figure 2**.

In disease mapping, these advances in remote sensing and sensor technology meant that identification of spatial heterogeneities would be possible even at small geographic or local scales. These advances were pivotal for disease mapping as epidemiologist could identify important drivers of disease risk and thus be able to guide control programs more efficiently and with evidence-based decisions. Also, the costs of remotely sensed data products had been significantly reduced

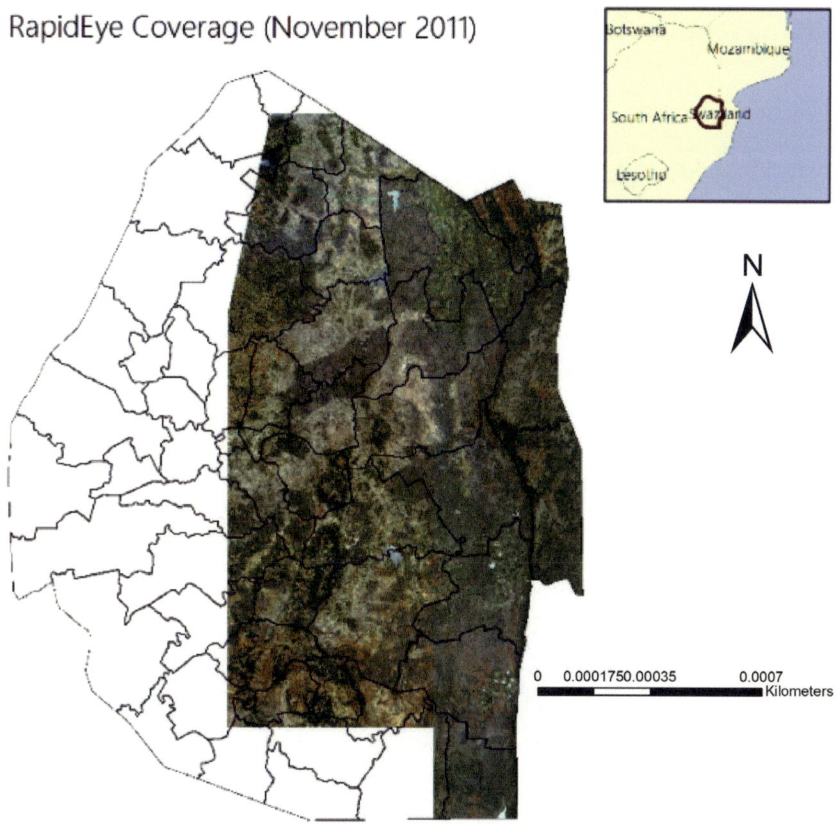

Figure 1.
An example of high-resolution true color image covering part of Eswatini used to develop a land cover classification.

as more sensors had been launched thus stabilizing the demand for data products and reliance in only a few remote sensing agents. More and more countries and private companies have launched satellites into space in the twenty-first century and the resultant imagery data products have been availed to the research community [24]. In addition, archived remotely sensed data products had often been offered to researchers free of charge and this have enabled spatial analysts to perform various analysis techniques such as time-series analysis, data mining and other data learning techniques. The RS data and other end products had also been customized in terms of the derivation and calculation of vegetation indices used in mapping studies. This customization had enabled direct incorporation of such indices into models as interpretation became possible to the research community even though not being experts in the remote sensing field. Common indices that had been widely adopted into disease mapping models because of their ease in interpretation include those of the Normalized Difference Vegetation Index (NDVI), temperature and rainfall. These indices had often been supplied or archived in their complete processed (derived) and customized form, thus enabling researchers to easily access them from the hosting agencies and websites and directly incorporate them into their mapping models as they had become interoperable.

Advances made in mapping software, particularly geographic information system (GIS) software, had seen interest being stimulated among disease researchers and epidemiologist. Whereas, earlier software was mostly geared toward solely remote sensing experts, the availability of customized mapping platforms for spatial epidemiology meant that these software programs could be utilized even by

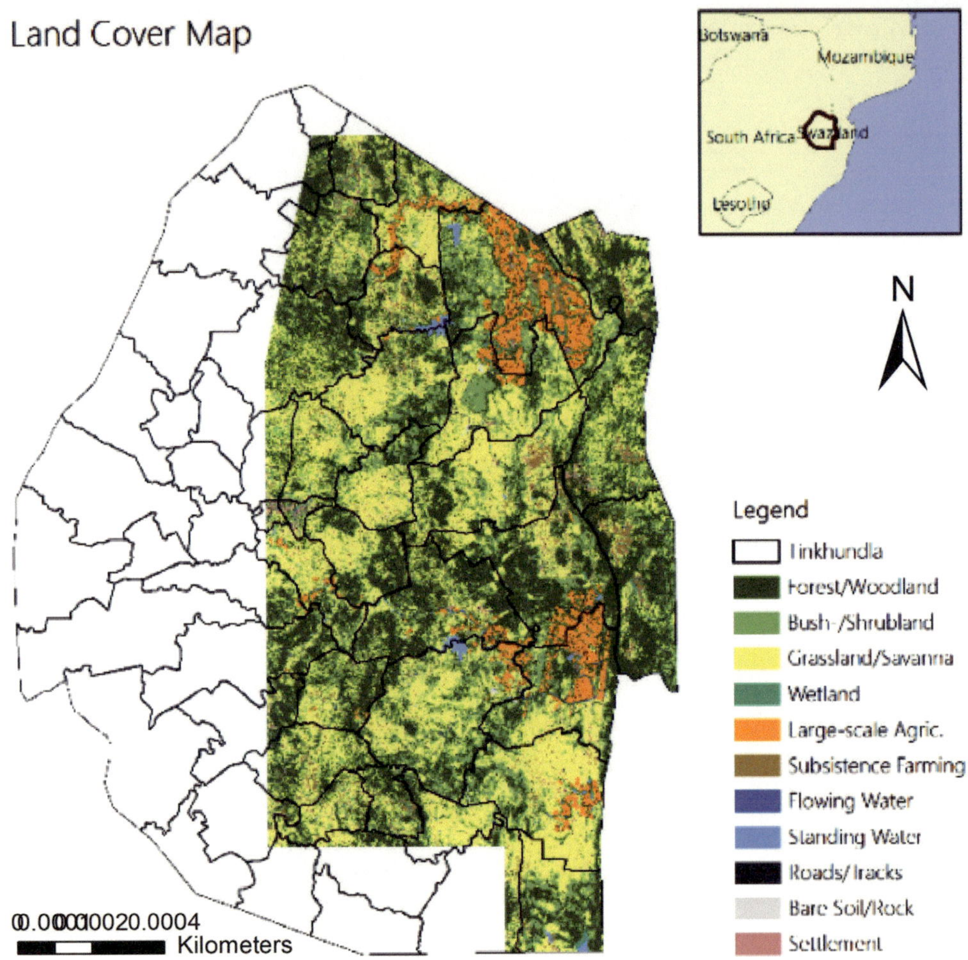

Figure 2.
An example of high-resolution land cover classification in Eswatini.

non-remote sensing experts. For instance, the public health community had been able to facilitate the development and customization of disease mapping software programs like; Health Mapper, Epi-Info, etc. Other GIS software programs such as ESRI ArcGIS had over the years added more customized mapping tools and extensions meant to support disease mapping and epidemiology efforts. Again, the high costs which were often associated with some of the commercial software had been reduced as more open source software became available. For example, GIS software programs such as QGIS had been availed as open source and could be directly downloaded and installed into any GIS capable computer. Also, true color visualization web-based software such as Google Earth which had even given mapping novices some level of confidence due to the fact that it had been made without any associated sophistication had contributed to the hype about disease mapping, rapid risk assessment, and prediction among epidemiologists. No mapping experts could leverage on such web-based imagery software programs and be able to identify, analyze, and interpret spatial phenomenon explicitly as it appears on the zoomed imagery on a computer screen. This way, disease experts had been able to explain some of the identified trends and patterns and also directly answer some of the pertinent questions associated with disease epidemiology such as clustering, severity variation and disease presence or absence *inter alia*.

Again, a number of statistical software programs have also added mapping extensions which have enabled the analysis of environmental and climatic data to be undertaken using such software. This has resulted to a new field of research called geostatistics, which combines both geography and statistics during spatial analysis. For example, statistical software programs such as STATA, R, WINBUGS, and others have been used to process and analyze climatic data derived from remote sensing. In disease risk mapping, space and time analysis had often been conducted using these statistical software programs and they had been widely used as research methods in epidemiology and disease risk prediction in addition to the usage of mathematical models which attempt to explain the underlying factors and quantities in disease risk modeling. Geostatistics therefore had been pivotal in the application and incorporation of remotely sensed data products into disease mapping and epidemiology. The capability of Geostatistics to incorporate technical algorithms that could be used to forecast disease burden in space and time had also contributed to the wide adoption of such approaches as it meant that control programs could *a priori* be informed about disease risk and thus be better prepared to deal with disease outbreaks.

As already mentioned, advances in computer processing power had enabled the integration of computing methods based for instance on Bayesian inference approaches which had been previously limited due to poor computer performance. Data simulation methods such as Markov Chain Monte Carlo (MCMC) and the integrated Laplace approximation (INLA) had been widely used to estimate posterior distribution of geographic data in space and time. The results of which had been obtainable within reasonable time frames compared to earlier computation efforts of similar data. In statistics MCMC are methods comprised of a class of algorithms that sample from a probability distribution. MCMC uses simulation techniques to find a posterior distribution and sample from it. On the other hand INLA relies on analytical combinations that approximate and efficiently integrate numerical schemes to achieve highly accurate deterministic approximations of posterior quantities of interest [25, 26]. As a result of these statistical and computational advances, the integration of environmental and climatic data derived from remote sensing technology into disease mapping models had over the years markedly increased.

Recently, the capability of big data, machine learning and other location intelligence methods to handle a large array of data sets have contributed to the awareness about the application of geographic data as often models using these methods would be performed on geographic software. Big data refers to extremely large data sets that may be computationally analyzed to reveal patterns, trends, and associations relating to human behavior. Machine learning approaches focuses on computer programs that can assess data and use them to learn and improve from experience of them without being explicitly programmed [27]. These analysis methods are often conducted on GIS capable computers and they also rely on remote sensing products such as satellite imagery to analyze and reveal any patterns, trends and associations coming from the data. In disease mapping and epidemiology such analysis approaches are important in understanding risk of disease spread due to human behavior and their interaction with the environment. As a result, most of the diseases mapping efforts currently applied have either practically or theoretically ended up either utilizing remotely sensed data or its associated geographic data products into spatial analysis models as often the resultant modeling outputs would be displayed in a mapping environment.

2.1 Remote sensing data application in vector-borne disease surveillance

Mapping of vector-borne diseases began around 1950 with the use of aerial photography and cartographic techniques. Early studies included those that focused on eradicating malaria, dengue fever and yellow fever whereby climatic factors were used to identify areas at risk of higher transmission. The Malaria Atlas Project founded in 2006 took over from previous mapping efforts and demonstrated the application of geography based variables to map and disseminate accurate information on malaria endemicity. Identifying and mapping vector habitats using climatic suitability was used to guide surveillance and control efforts. Different approaches were used to improve visualization and to produce detailed maps such as high-altitude color-infrared photography and also incorporating high-resolution images [28]. Mapping of vegetation types associated with some of the vector breeding habitat had been carried out since 1973. The techniques used in such analysis had been very important for surveillance support and for identification of vector oviposition habitats. The visualization and interpretation techniques used were based on tone and texture and were used to identify habitats associated with tick-borne disease in some areas based on the concepts of landscape epidemiology of disease [29].

In early 1970s, multispectral scanner data was first used to monitor and map environmental parameters required for the breeding of disease vectors. A combination of remote sensing data acquired from satellites as well as aircraft platforms was used for this task. Later in 1976, some studies demonstrated that computer processing techniques could be used to classify airborne multispectral scanner data for mapping and identifying vegetation types associated with certain disease-causing mosquitoes. Around 1984, remote sensing techniques were applied to describe and map geographic characteristics associated with schistosomiasis [30]. Temperature and precipitation data obtained from remote sensing was also used to estimate the probability of disease occurrence at unsampled locations. These data were also used to identify and map mosquito larval habitats and their association with certain environmental variables in space and time. A study by [31] identified tick habitats on the island of Guadeloupe using derived vegetation and moisture indices.

Most of the studies discussed above were primarily focused on the application of remote sensing to identify and map potential vector habitats and breeding sites based on vegetation, water, and soil. Identifying existing or potential habitats and breeding sites would not be enough to adequately guide surveillance and control efforts unless all possible affected areas were identified and mapped. Therefore most studies have also incorporated predictive techniques as part of their support of surveillance and control efforts. Consequently, it had been necessary for studies to go beyond mere habitat and potential breeding sites mapping and to make predictions of vector distributions in space and time often for the entire geographic area of interest. This often included identification and mapping areas where vector production and disease transmission risk would be greatest in defined time and space thresholds. **Figure 3** is an example of the spatial distribution of malaria vector breeding sites and their distance to subsistence farming in Eswatini.

2.2 Predicting diseases using remotely sensed data variables

An important aspect in the application of RS data in disease mapping and epidemiology is their use as predictor variables in modeling. Disease mapping studies had often used environmental and climatic proxies derived from remote sensing in statistical regression models aiming to predict disease risk in both its spatial and temporal dynamics. These studies predict disease distributions, vector populations and disease transmission risk within the affected populations in specific areas. Common climatic variables used in disease predictive modeling studies often combine remote sensing measurements of vegetation, precipitation, and temperature to identify when and where conditions would be favorable for disease propagation. Other studies have attempted to use remote sensing to predict the temporal as well

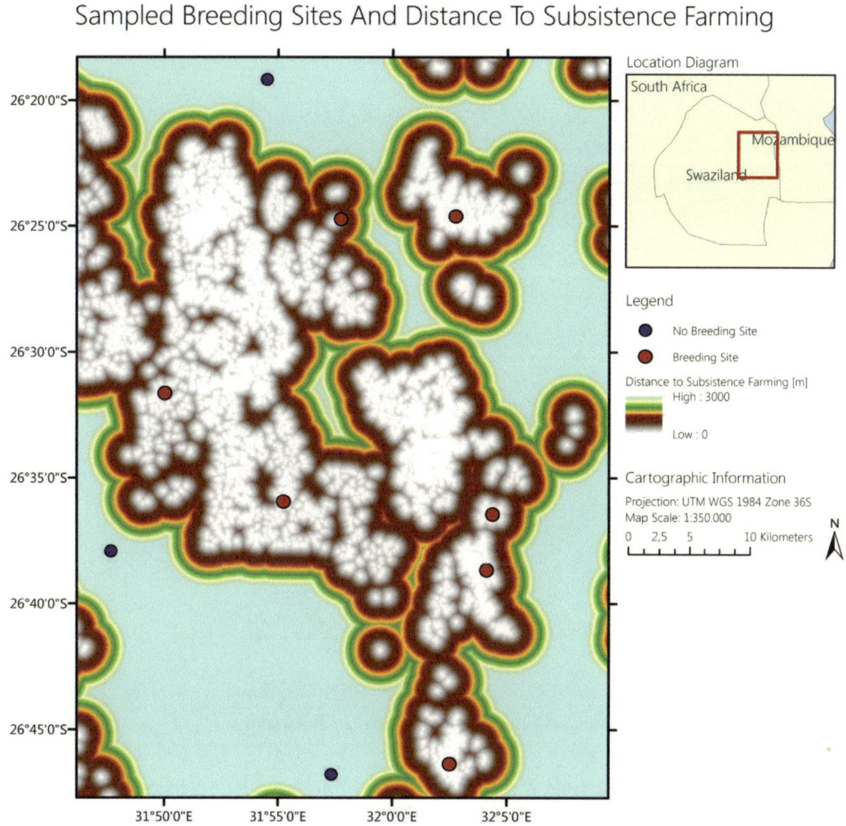

Figure 3.
Example of mapped breeding sites and their distance to subsistence farming.

as spatial patterns, of habitat development, vector populations, and disease transmission risk [32, 33].

Among the most common disease mapping studies where predictions had often been used included those of malaria. These studies also use remote sensing to predict which populations or villages are at risk of transmission. In these cases risk would often be defined by the proximity of a village to areas of heightened transmission as well as the breeding, feeding, and resting habitats required by the malaria vector *Anopheles* mosquito. In addition recent studies have focused on assessing the issue of predictive modeling and disease transmission risk based on the application of remote sensing and GIS technologies. To date, there have not yet been an alternative to disease mapping and prediction in space and time and to identify, characterize, and map the patterns of vector habitats other than using products derived from satellite imagery and remote sensing. For instance, identification and location of areas where vector survival rates are highest had been done using vegetation indices derived from remote sensing. **Figure 4** is an example of predicted potential malaria vector breeding sites in the northeastern part of Eswatini.

Also remote sensing data and GIS techniques had been used to identify and map landscape features associated with disease transmission risk. Landscape features such as brush, woodland and grassland and areas cleared for housing, roads, or trails and other similar locations identified through remote sensing had often been used for detecting intercept areas between human hosts, vectors and parasites. Other landscape features such as coniferous forest, deciduous forest, mixed forest,

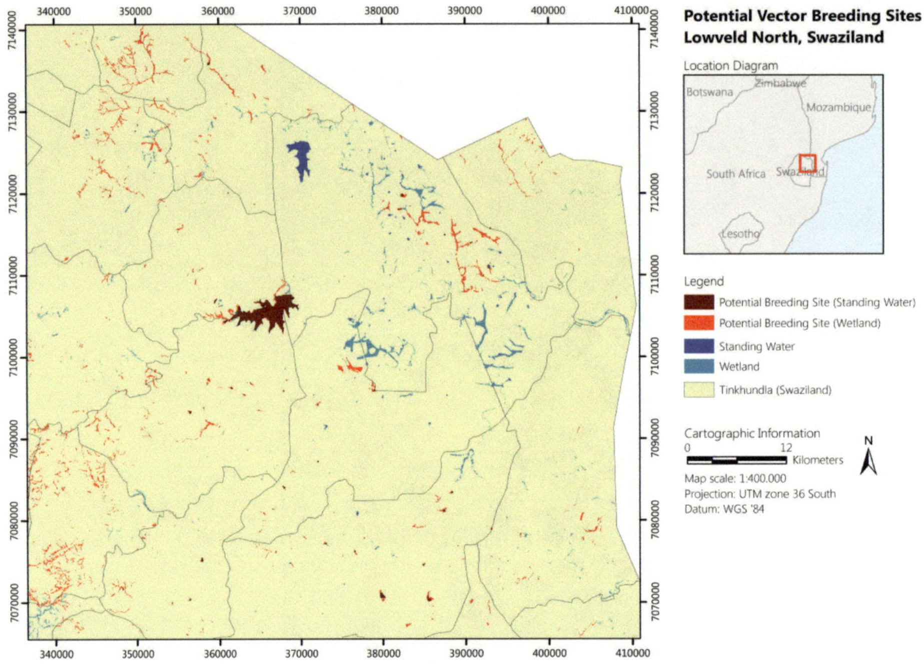

Figure 4.
An example of identified potential malaria vector breeding sites in Eswatini.

water bodies, glades, and housing developments had also been identified via remote sensing. Some studies have assessed these landscapes for their association with the presence of certain disease vectors and the proximity of housing as a measure of human exposure to those diseases. Such landscape epidemiology had therefore been used to identify areas where transmission risk is greatest by characterizing the mixture of deciduous forest and residential developments that bring diseases, vectors, and humans into contact.

Field studies had often been used to train spatial models that make use of remote sensing to map the distribution of disease and predict transmission risk areas. These studies also use GIS to capture groundtruthed coordinates or Global Positioning System point (GPS) which would then be used to assess the accuracy of predictive models. For instance, GIS and remote sensing had been used to investigate the adjacency of certain landscape features and residential properties with dense vegetation as a potential measure of human-vector contact. In this case, regression models would be used to assess general correlations between landscape and disease transmission risk. Evidently, remote sensing and GIS had been combined to study, for instance, the structure and composition of a landscape as it relates to the epidemiology of a disease. Again some studies have combined remote sensing and GIS analysis techniques, to assess various associations between diseases, vectors and human contacts.

Models based on remote sensing data and GIS techniques have also been used to study certain disease vector population dynamics using remote sensing and GIS technologies. Such studies use satellite imagery and GIS modeling techniques to distinguish between areas with either high or low disease-causing vectors. In these cases, ground data on vector populations are used jointly with remote sensing data combined via either a statistical or a GIS software. Ground data variables on vector presence or absence are analyzed in relation to the remotely-sensed spectral data captured via satellite imagery. The groundtruthed field measurement data, that were observable using GPS, would then be used to evaluate the accuracy of the remote sensing based predictive models.

2.3 Current applications of remote sensing in epidemiology

Current research is focusing on the capabilities of remote sensing and GIS data to perform various spatial analyses. The goal is to describe the landscape, or land cover, composition that explains better the distribution of diseases. In this regard spatial models mounted on different software platforms are being developed and validated with observed data to predict spatial phenomenon and in particular diseases that are of public concern. The approaches used to process, analyze and fit data into models are also constantly evolving as new software programs and tool extensions and functionalities become available. There is also an increase in exploratory research involving mathematical and statistical models which aim to capture both the deterministic and the stochastic components during data analyses. In such cases dynamic models involving Susceptible Infectious and Recovered models (SIR) used to model highly infectious diseases such as COVID-19 are being extended by adding probabilistic based components in order to model uncertainty in the behavior of the diseases among the affected population. In these instances, environmentally determined infectious diseases rely on climatic and weather data derived from remote sensing for effective modeling. Furthermore, Satellite imagery could be used to analyze the socioeconomic changes that are currently taking place as a result of COVID-19. This could also be useful in identifying the impacts that measurers such as the national lockdowns are having on the environment. Depending on the resolution of the RS data product being utilized, frequency of data capture and timeliness of the data capture, there is a high potential for remote sensing to be used a tool for deployment and monitoring the effects of health interventions implemented to fight COVID-19.

Point process models such as Log-Gaussian Cox Process had been developed are also being used to model climatic and environmental data on fine geographic scale. These models combine a Poisson process in the first level with a Gaussian Process at the second level and are used to analyze point patterns. In such cases very high-resolution remotely sensed data would be used to enhance boundary delineation during mapping especially at local scales. Tools like spatial scan statistics had also been used to identify and map disease clusters and to determine the key driving factors resulting in the identified clusters. Spatial scan statistics defined as the maximum likelihood ratio statistics over a collection of scanning windows had also been widely used to determine clustering in space and time. Research studies applying remote sensing in disease mapping and epidemiology are currently being undertaken at various degrees of complexity as new methods and techniques become available. The primary focus had been on the use of remote sensing and GIS capabilities to quantify various disease determining factors and to estimate the probability that vector-borne diseases will be more abundant in some of the identified habitats as well as to determine the factors necessary for transmission, survival and reproduction.

3. Conclusions

As already discussed, over the past 30 years, prolific research studies demonstrating the potential opportunities offered by the use of remote sensing, GIS, and statistics in disease mapping and epidemiology had been undertaken. Most of such studies aimed to analyze and measure some of the presumed associations between environmental factors and human diseases. The results of the studies mentioned above had been used to guide the public health community during health intervention planning and decision-making. The studies had also been used to demonstrate the potential offered by remote sensing application to disease mapping and epidemiology and to support surveillance and control efforts. They have illustrated the diversity of potential remote sensing applications in disease surveillance and control programs. However, successful application of remote sensing technology depends on the ability of nonexperts' RS data and end users to access, retrieve, and analyze the data captured from satellites. The preprocessing steps involved before such data could be added as covariates into models also determine their uptake and usage by nonexperts. The ability to develop near real-time monitoring spatial models in order to timely predict the spatial and temporal patterns of vector-borne diseases, and transmission risk is also a motivation for the use of RS data in disease mapping and epidemiology.

Clearly, the dynamics of vector-borne diseases at any location are influenced by processes that operate at a variety of landscape and geographic scales. For instance, malaria transmission is a result of spatial interaction between hosts, vectors, and parasites. Remote sensing imagery involving both high and coarse resolutions when combined with GIS spatial analyses techniques could be used to support and guide existing vector surveillance and control programs at local, regional, and even continental scales. The findings of some the studies cited above had been used to illustrate and cement how remote sensing and GIS technologies can provide epidemiologists with a new perspective in as far as determining the environmental drivers of the diseases concerned. Researchers had been able to study the multiple factors influencing the patterns and distributions of vector-borne diseases at a variety of landscape and geographic scales. The exploration of some of the opportunities presented by remote sensing to disease mapping and epidemiology is still unfolding as new opportunities are being presented.

Acknowledgements

The author would like to thank the University of Eswatini for promoting and supporting the ongoing scientific research and publication.

Conflict of interest

The author declares no conflict of interest.

Thanks

The author would like to thank his family for the support they gave during the write-up of this chapter, which also coincided with the nationwide lockdown implemented to curb COVID-19 in Eswatini.

Author details

Sabelo Nick Dlamini
University of Eswatini, Manzini, Eswatini

*Address all correspondence to: sabelodlamini2004@yahoo.com

References

[1] Campbell JB, Wynne RH. Introduction to Remote Sensing. 5th ed. New York, NY: Guilford Press; 2011. p. 717

[2] Viña A, Gitelson AA, Nguy-Robertson AL, Peng Y. Comparison of different vegetation indices for the remote assessment of green leaf area index of crops. Remote Sensing of Environment. 2011;**115**(12):3468-3478

[3] Mwakapuja F, Liwa E, Kashaigili J. Usage of Indices for Extraction of Built-up Areas and Vegetation Features from Landsat TM Image: A Case of Dar Es Salaam and Kisarawe Peri-Urban Areas, Tanzania | Francis Mwakapuja - Academia.edu [Internet]. 2013. Available from: http://www.academia.edu/9341512/Usage_of_Indices_for_Extraction_of_Built-up_Areas_and_Vegetation_Features_from_Landsat_TM_Image_A_Case_of_Dar_Es_Salaam_and_Kisarawe_Peri-Urban_Areas_Tanzania [Accessed: 26 November 2018]

[4] Noor AM, Kinyoki DK, Mundia CW, Kabaria CW, Mutua JW, Alegana VA, et al. The changing risk of plasmodium falciparum malaria infection in Africa: 2000-10: A spatial and temporal analysis of transmission intensity. Lancet. 2014;**383**(9930):1739-1747

[5] Karagiannis-Voules D-A, Biedermann P, Ekpo UF, Garba A, Langer E, Mathieu E, et al. Spatial and temporal distribution of soil-transmitted helminth infection in sub-Saharan Africa: A systematic review and geostatistical meta-analysis. The Lancet Infectious Diseases. 2015;**15**(1):74-84

[6] Karagiannis-Voules D-A, Odermatt P, Biedermann P, Khieu V, Schär F, Muth S, et al. Geostatistical modelling of soil-transmitted helminth infection in Cambodia: do socioeconomic factors improve predictions? Acta Tropica. 2015;**141**(Pt B):204-212

[7] Clements ACA, Kur LW, Gatpan G, Ngondi JM, Emerson PM, Lado M, et al. Targeting trachoma control through risk mapping: The example of southern Sudan. PLoS Neglected Tropical Diseases. 2010;**4**(8):e799

[8] Robinson T, Rogers D, Williams B. Mapping tsetse habitat suitability in the common fly belt of Southern Africa using multivariate analysis of climate and remotely sensed vegetation data. Medical and Veterinary Entomology. 1997;**11**(3):235-245

[9] Xu M, Cao C, Wang D, Kan B. Identifying environmental risk factors of cholera in a coastal area with geospatial technologies. International Journal of Environmental Research and Public Health. 29 Dec 2014;**12**(1):354-370. DOI: 10.3390/ijerph120100354. PMID: 25551518. PMCID: PMC4306866

[10] Karume K, Schmidt C, Kundert K, Bagula ME, Safina BF, Schomacker R, et al. Use of remote sensing for population number determination. The Open Access Journal of Science and Technology [Internet]. 2017;**5**(03):9. Available from: https://www.agialpress.com/articles/use-of-remote-sensing-for-population-number-determination.pdf [Accessed: 26 November 2018]

[11] Yang G-J, Vounatsou P, Xiao-Nong Z, Utzinger J, Tanner M. A review of geographic information system and remote sensing with applications to the epidemiology and control of schistosomiasis in China. Acta Tropica. 2005;**96**(2):117-129

[12] Kaptein A, Janoth J, Lang O, Bernede N. Trends in commercial radar remote sensing industry [industrial profiles]. IEEE Geoscience and Remote Sensing Magazine. 2014;**2**(1):42-46

[13] Correia VR de M, Carvalho MS, Sabroza PC, Vasconcelos CH. Remote sensing as a tool to survey endemic diseases in Brazil. Cadernos de Saúde Pública. 2004;**20**(4):891-904

[14] Thomson MC, Connor SJ, Milligan P, Flasse SP. Mapping malaria risk in Africa: What can satellite data contribute? Parasitology Today. 1997;**13**(8):313-318

[15] Beck LR, Rodriguez MH, Dister SW, Rodriguez AD, Rejmankova E, Ulloa A, et al. Remote sensing as a landscape epidemiologic tool to identify villages at high risk for malaria transmission. The American Journal of Tropical Medicine and Hygiene. 1994;**51**(3):271-280

[16] Hay SI, Packer MJ, Rogers DJ. The impact of remote sensing on the study and control of invertebrate intermediate hosts and vectors for disease. International Journal of Remote Sensing. 1997;**18**(14):2899-2930

[17] Chilès J-P, Delfiner P. Geostatistics: Modeling Spatial Uncertainty. Hoboken, New Jersey: John Wiley & Sons, Inc.; 2009. p. 718

[18] Tran A, Ippoliti C, Balenghien T, Conte A, Gely M, Calistri P, et al. A geographical information system-based multicriteria evaluation to map areas at risk for Rift Valley fever vector-borne transmission in Italy. Transboundary and Emerging Diseases. 2013;**60** (Suppl 2):14-23

[19] Hassan AN, Beck LR, Dister S. Prediction of villages at risk for filariasis transmission in the Nile Delta using remote sensing and geographic information system technologies. Journal of the Egyptian Society of Parasitology. 1998;**28**(1):75-87

[20] Dlamini SN, Franke J, Vounatsou P. Assessing the relationship between environmental factors and malaria vector breeding sites in Swaziland using multi-scale remotely sensed data. Geospatial Health. 2015;**10**(1):302

[21] Garni R, Tran A, Guis H, Baldet T, Benallal K, Boubidi S, et al. Remote sensing, land cover changes, and vector-borne diseases: Use of high spatial resolution satellite imagery to map the risk of occurrence of cutaneous leishmaniasis in Ghardaia, Algeria. Infection, Genetics and Evolution. 2014;**28**:725-734

[22] Sadeghieh T, Waddell LA, Ng V, Hall A, Sargeant J. A scoping review of importation and predictive models related to vector-borne diseases, pathogens, reservoirs, or vectors (1999-2016). PLoS One. 2020;**15**(1):e0227678

[23] Leonardo LR, Rivera PT, Crisostomo BA, Sarol JN, Bantayan NC, Tiu WU, et al. A study of the environmental determinants of malaria and schistosomiasis in the Philippines using remote sensing and geographic information systems. Parassitologia. 2005;**47**(1):105-114

[24] Dlamini SN, Beloconi A, Mabaso S, Vounatsou P, Impouma B, Fall IS. Review of remotely sensed data products for disease mapping and epidemiology. Remote Sensing Applications: Society and Environment. 2019;**14**:108-118

[25] Geyer CJ. Practical Markov Chain Monte Carlo. Statistical Science. 1992;**7**(4):473-483

[26] Illian JB, Sørbye SH, Rue H. A toolbox for fitting complex spatial point process models using integrated nested Laplace approximation (INLA). The Annals of Applied Statistics. 2012;**6**(4):1499-1530

[27] Beam AL, Kohane IS. Big data and machine learning in health care. Journal of the American Medical Association. 2018;**319**(13):1317-1318

[28] Wagner VE, Hill-rowley R, Narlock SA, Newson HD. Remote sensing: A rapid and accurate method of data acquisition for a newly formed mosquito control district. Mosquito News. 1979;**39**(2):283-287

[29] Pavlovsky EN. Natural Nidality of Transmissible Diseases with Special Reference to the Landscape Epidemiology of Zooanthroponoses [Internet]. 1966. Available from: https://www.cabdirect.org/cabdirect/abstract/19672901301 [Accessed: 31 July 2020]

[30] Cross ER, Perrine R, Sheffield C, Pazzaglia G. Predicting areas endemic for Schistosomiasis using weather variables and a Landsat Data Base. Military Medicine. 1984;**149**(10):542-544

[31] Hugh-Jones M, Barre N, Nelson G, Wehnes K, Warner J, Garvin J, et al. Landsat-TM identification of Amblyomma variegatum (Acari: Ixodidae) habitats in Guadeloupe. Remote Sensing of Environment. 1992;**40**(1):43-55

[32] Linthicum KJ, Bailey CL, Davies FG, Tucker CJ. Detection of Rift Valley fever viral activity in Kenya by satellite remote sensing imagery. Science. 1987;**235**(4796):1656-1659

[33] Pope KO, Sheffner EJ, Linthicum KJ, Bailey CL, Logan TM, Kasischke ES, et al. Identification of central Kenyan Rift Valley fever virus vector habitats with landsat TM and evaluation of their flooding status with airborne imaging radar. Remote Sensing of Environment. 1992;**40**(3):185-196